D0722846

ECOTOPIA 2121

A VISION FOR OUR FUTURE GREEN UTOPIA—IN 100 CITIES

DR. ALAN MARSHALL

Arcade Publishing ¤ New York

First Edition

Arcade Publishing books may be purchased in bulk at special discounts for sales promotion, corporate gifts, fund-raising, or educational purposes. Special editions can also be created to specifications. For details, contact the Special Sales Department, Arcade Publishing, 307 West 36th Street, 11th Floor, New York, NY 10018 or arcade@skyhorsepublishing.com.

Arcade Publishing® is a registered trademark of Skyhorse Publishing, Inc.®, a Delaware corporation.

Visit our website at www.arcadepub.com.
Visit the author's website at www.ecotopia2121.com.

10 9 8 7 6 5 4 3 2 1

Names: Marshall, Alan, 1969– author.
Title: Ecotopia 2121 : a vision for our future green utopia—in 100 cities /
    Alan Marshall.
Description: New York : Arcade Publishing, 2016. | Includes bibliographical
    references and index.
Identifiers: LCCN 2016018068 (print) | LCCN 2016021075 (ebook) | ISBN
    9781628726008 (hardback) | ISBN 9781628726145 (ebook)
Subjects: LCSH: Sustainable urban development. | Urban ecology (Sociology) |
    City planning. | BISAC: NATURE / Ecology. | SOCIAL SCIENCE / Sociology /
    Urban. | ARCHITECTURE / Urban & Land Use Planning.
Classification: LCC HT241 .M37 2016 (print) | LCC HT241 (ebook) | DDC
    307.1/16—dc23
LC record available at https://lccn.loc.gov/2016018068

Cover design by Laura Klynstra
Cover illustration: Vienna 2121 by Diversepixel. Used by permission.

Printed in China

Contents

Contents

¤ ¤ ¤

Contents

Contents

¤ ¤ ¤

Contents

□ □ □

Contents

¤ ¤ ¤

THE CITIES OF ECOTOPIA

Introduction

Utopia is that place on Earth—real or imagined—where society is resplendent with all things harmonious and good, and the people who live there are free and happy. Utopian cities have appeared in fables and stories, on film, and in manifestos. Perhaps, also, utopian cities have existed in the real world, if only in localized places and for a short time.

Sometimes the word *utopia* is bandied about pejoratively as referring to a place in a fantasy world. In this way of thinking, "utopia" is the label some use to describe a place where we all might desire to live but which is ultimately unattainable. Equally, though, the word is often used in a serious and earnest manner to mean that, through radical change or gradual evolution, the world can develop into an ideal, even idyllic, place. This book might be seen as sympathetic to the latter way of thinking—even though the "serious" tag may not apply to each and every page.

Utopia is sometimes theorized based on a single relevant feature—for example, the abolition of all human suffering or the achievement of full equality. But it is just as valid to posit it in a multidimensional fashion involving many features. Some commonly expressed themes in utopian writing include the following scenarios:

- Social harmony and justice meld symbiotically with personal freedom.
- Peace and love overcome war and hatred.
- Material well-being is achieved for all people without recourse to greed or exploitation.
- Work, if it exists, is almost always enjoyable and human life is almost always satisfying.
- Equality between people, in one form or another, is fostered and honored, along with the rights of each individual.

Whatever particular set of features is highlighted by any one utopian idealist, all utopians convey the theme that utopia is not a paradise created by divine intervention or by some supernatural process but is the product of the combined actions of humans (who've somehow stumbled upon a good recipe for social organization).

Utopia is a centuries-old concept, of course. Thomas More, an English cleric and statesman, first released the term upon the world in late 1516 when he published his book *Utopia*, about a mysterious and near-perfect island state. At the time, More was reflecting upon the tales brought back from the New World by European explorers like Christopher Columbus and Amerigo Vespucci, including tales about American "Indians." In a way, his book, *Utopia*, is an attempt to make sense of these stories of indigenous societies.

More was being intentionally ambiguous when he invented the word *utopia*. It is derived from the Greek language, but in the process of importing it, More combined two possible meanings. His word could refer to either a *good land* or a *nowhere land*, or, indeed, both. This is the case because, while the Greek *topos* translates to "place," the "u" part could come from the Greek ε□, which means "good," or from the Greek o□, which means "not,"

VTOPIAE INSVLAE FIGVRA

or "none," or "nothing." Did Thomas More mean to say that utopia is a *good land* or a *land that does not exist*? Probably he meant both.

To design a utopian land often means to be at once imaginative and optimistic but also critical and subversive. Thomas More set up this enduring pattern when he created his optimistic account of an idealistic utopia within which was embedded a subtle and subversive critique of King Henry VIII's reign in England. In the end, Henry VIII had More's head chopped off and set on a pike over London Bridge. This wasn't in retaliation for writing *Utopia* specifically, but for siding with the Catholic Church against Henry VIII as the king turned his back on Rome to seek a divorce from Catherine of Aragon. For his unabashed commitment to Roman Catholicism, Thomas More was declared the patron saint of politicians centuries later.

Strangely, perhaps, More was also honored by the Soviet Union because of his communistic idea about the common ownership of property. In Soviet Russia, the name of Thomas More was in ninth position from the top of Moscow's Obelisk of Revolutionary Thinkers. The obelisk was removed at the beginning of the Putin era—some say because the modern rulers of Russia did not want any more revolutions, and others say because the modern rulers of Russia did not want any more thinkers.

After Thomas More's book, whenever utopian writers have set out to design or discover a socially ideal place, they've also used the occasion to attack certain aspects of their own present-day society—cloaking their visions in both hope and satire. I dare say this may be the same impulse that flows through many of the utopian scenarios presented in the chapters that follow, as I seek to imagine utopian cities approximately one hundred years hence. As I do this, I'm hopeful my own head won't end up on a pike. However, it should be acknowledged that within many of the cities explored here, especially those without democratic rule, any public call for alternative forms of governance is either frowned upon or, indeed, infringes upon one or another law. In the course of my research, I've visited many of the cities included here—and I've lived and worked in quite a few of them, too—but I'm sure the reactionaries in many of them would not invite me back should they chance upon some of the scenarios expressed in these pages.

◻ ◻ ◻

Cities are where most humans live out their lives. At the moment, around 55 percent of the world's people live in cities, and by the end of the century

Introduction

◻ ◻ ◻

Introduction

⌑ ⌑ ⌑

the proportion is likely to be much greater, maybe reaching some three-quarters or more. Hence, the importance of cities to the future of humanity and the world can hardly be overestimated.

Cities have been facing myriad problems since they first evolved some four thousand years ago. As well as perennial issues like resource scarcity, sanitation, and crime, nowadays almost all cities have to face up to the global environmental crisis. For this reason, utopian thinkers these days are at times keen to add ideas about ecological harmony to those of social harmony. This type of utopia has been called a *Green utopia* or *ecotopia* by a number of authors. Ecotopian writers fantasize just as much about living in a clean and green environment as they do about freedom, happiness, and social equality. For some, indeed, the former is more important than the latter. This book attempts to describe a world where these Green dreams have become manifest by the year 2121.

Many ecotopian thinkers would probably accept the basic social features of utopia as cited here—freedom, equality, and the desire for happiness—but they would say that ecological principles must also be valued and adhered to. Thus, our liberties

and our material creations should be enjoyed without harming the natural environment. In an ecotopia, the relationship of society to nature is changed so that humans act fundamentally to preserve or defend some facet of their natural setting, be that their local environment or the entire global biosphere.

According to the logic of environmentalism, utopia is sustainable only if the well-being of the world's ecology is part of the plan. And so, imagined herein are the various ecotopian futures of one hundred cities across the globe. They reflect a wide range of geographies and cultures over diverse social settings. Each of the scenarios contains a description of what the city's proposed future consists of, and also some explanation of how the city might be transformed from what it is today to how it is projected to be some one hundred years hence.

The future of the chosen cities is expressed in the form of *scenario art*, presented with one piece of artwork and a few pages of narrative text. Scenario art is sometimes utilized as a way to get people to seriously envisage alternative future plans. I take note of this way of imagining, debating, and communicating decisions but acknowledge that art—in theory and in practice—has often reveled in intense self-reflection, as well as speculation, over and above any practical factors.

The one hundred pieces of art depicting the one hundred different cities follow sometimes quite different styles, as does the narration about the cities. Some of these styles tend toward the abstract and intellectual, but other scenarios are far more personal and tied to everyday events, including those in my own life. I'm hoping they will be viewed in concert as a symphony of human responses to particular urban settings, exposing—rather tangentially sometimes—diverse dreams of a better future.

One problem I foresee for certain readers is that some may get a little uptight about whether I am trying to be realistic or not. On this point, I must convey the following. This work is a journey through space and time. In the pages that follow, we roam across the lands of the Earth, to all seven continents, and we roam back and forth across time, from the distant pasts of cities to our own time, and to the century ahead. From a geographical perspective, all the cities presented in this book are based on real cities, not imagined places, barring one exception. From a temporal

perspective, the stories I've taken from history are those expressed in shared historical narratives. They are not fictitious—again, barring one exception. Likewise, the contemporary descriptions of the one hundred cities, such as they exist today, are not my own imaginings but are distilled from both my own research and the research of others. As for the future scenarios, imagination and speculation are involved, to be sure, and at times this journey into the future will get a little wild for some readers, though even the wildest futures foretold here are heavily reinforced with scholarly exploration.

Although many social studies academics have a disaffection for making projections about the future, most of them would admit that one thing is certain: eventually, the twenty-second century will roll around, and at that time human beings will still probably be living together in cities. It seems to me that it is not a waste of time to think about this future, even though readers will not likely be around to see it, for in exploring our ideas of future cities, we bring into sharp focus our own assumptions about better tomorrows.

At some point while perusing this work, the reader is bound to ask why the year 2121 has been chosen as the scenario date. The answer is simple.

Introduction

Around the world, planners have adopted *2020 vision* statements for specific cities because they think it cutely allies the metaphor of *20/20 vision* with 2020 AD, a date in the future they think they have some identification with or some influence over. For 2121, our vision is not perfect and our influence may be rather tenuous. But, because 2121 is somewhat fuzzy and uncertain, the need for us to construct a vision with the help of imagination is that much greater.

Introduction

¤ ¤ ¤

ECOTOPIA 2121

Abu Dhabi 2121 ¤ In the Shade of a Palm

"Cities grow great only when citizens plant trees whose shade they will not live to enjoy." So says an ancient proverb shared by many peoples of the lands surrounding the Persian and Arabian Seas, including the Emirati Arabs. The capital city of the Emirati Arabs is Abu Dhabi, the neighboring brother-city of Dubai. Compared to many contemporary Arab cities, both Abu Dhabi and Dubai are the most ultramodern and wealthy. Their respective approaches to urban planning and architecture have been rather different, though. Whereas Abu Dhabi is mostly low-rise and functional, Dubai is adorned with hundreds of glorious towers and large-scale megaprojects, such as offshore luxury suburbs built on huge, artificial palm-shaped islands (boasted of as being so grand that they are visible to astronauts in space).

In 2010, the tallest building in the world, the Dubai Tower, was opened. Reaching nearly one kilometer, or more than twenty-seven hundred feet, into the desert air from the center of Dubai, it's a towering visual symbol of Dubai's prosperity. However, the finances never really added up. Shortly before the opening, the Dubai builders went bankrupt and had to be bailed out by King Khalifa, the ruler of Abu Dhabi. The Dubai Tower was then quickly renamed the Khalifa Tower in deference to this act of financial rescue.

For many years, Abu Dhabi usually looked on with dismissive bemusement rather than envy at the megaconstructions of its ostentatious brother city. Now, though, Abu Dhabi itself has started some grandiose projects, including the construction of the world's tallest flagpole, a series of one-hundred-floor skyscrapers, and a cohort of glamorous art galleries, museums, and universities.

These projects are being brought to fruition through the use of cheap labor from the Indian subcontinent. The laborers often suffer atrocious working conditions. They have to put up with long hours, low wages, oppressive visa regulations, and cramped living conditions in substandard housing. Usually, they are corralled from their dormitories at 6:00 a.m. onto a jam-packed company bus, taken to work on dangerous worksites, and then twelve hours later they are herded again onto the same bus and taken straight back to their dormitories. This daily routine is repeated six or seven days a week for years until a project is completed. Sometimes, though, a project is abandoned by a company and the laborers also end up abandoned, without compensation, without work

Abu Dhabi 2121

¤ ¤ ¤

3

permits, and without a way to get back to their home countries.

For decades, the Abu Dhabi and Dubai governments have justified their treatment of South Asian immigrant workers by declaring that the laborers are provided with an economic opportunity that is unavailable to them in their own nation. Immigrants make up some 80 percent of Abu Dhabi's population, but very few immigrants are ever granted Emirati citizenship, even if they have lived there for decades, so they never attain anything resembling equal rights with the local Arab Emiratis. As the early twenty-first century unfolds, if these workers continue to be treated so badly, then the tensions between the large immigrant majority and the local Emirati minority may very well result in an all-out revolution. The scenario depicted here of Abu Dhabi 2121 portrays the city many years after such a revolution, when society has fundamentally changed. Here, Indian workers long ago overpowered King Khalifa's Abu Dhabi government before announcing secession from the Emirates to set up their own independent democratic nation. Their first act of law was to grant full citizenship rights to immigrant workers.

In this future of Abu Dhabi 2121, a half-built skeleton of a tower, once destined to be a private villa in the sky for the city's Emirati elite, is redeveloped into a massive palm tree. The palm emerges from the Abu Dhabi city center to shade the citizens from the desert sunshine. Typically, on summer days in central Abu Dhabi, the streets swelter in body-sapping heat, but here the palm tree offers free citywide relief from the sun. At present, the megastructures of the Emirates, like the Khalifa Tower and the Dubai palm islands, are built as private commercial endeavors or as nationalistic monuments. The Abu Dhabi 2121 Palm, however, stands in contrast as a public good, acting as a passive cooling device to provide an eco-friendly alternative to the ubiquitous use of energy-greedy air conditioner units.

¤ ¤ ¤

In the early twenty-first century, Abu Dhabi is a car-dependent city on par with any in the Western world—a situation fueled by the Emirati desire to travel in air-conditioned luxury as well as by the availability of cheap indigenous oil. Here in Abu Dhabi 2121, however, the stranglehold of the car on city life is broken with the help of a new socio-architectural setting comprising one- or two-story dwellings interconnected with walkways. The dwellings, partly inspired by the domestic houses in the Thar Desert of India, are constructed from local sands and muds mixed with native palm leaves and dried camel dung. This technique makes for buildings that need less energy to construct but that also possess a high degree of insulation. The dwellings serve both as family homes and as small businesses, and their walkways connect the community while providing shade for pedestrians and gardens below. The noisy, dangerous Abu Dhabi inner suburbs of today are in this way converted into relaxed neighborhoods where people can walk conveniently on flat surfaces without the need to compete with or navigate around cars or an obstructive roadscape.

Today, in the early decades of the twenty-first century, the Abu Dhabi economy is largely based on oil. In the future, though, oil reserves will very possibly have drastically declined—perhaps hastening the decline of the private automobile as well. In this post–Peak Oil age, crude oil will probably be available only for the production of a few essential plastic goods and medicines, not for cars and transport. However, the Abu Dhabi economy can still thrive if the residents there can learn to harness the energy of that other, ever-present, renewable resource that the city enjoys: the desert sunshine.

Abu Dhabi 2121

¤ ¤ ¤

Accra 2121 ⋈ Rising above the Flood

The capital of Ghana is increasingly exposed to lethal and costly urban flooding, made worse by unregulated construction over waterways and streams and by rivers being clogged with garbage. If the floods continue into the future in this chronic manner, they will probably encourage the people of Accra's flood-prone zones to migrate to other areas.

Over the coming decades, those families who build their houses above flood level will avoid disaster. After one hundred years of this ongoing process, either the whole population of Accra will have migrated inland or they will have built their homes in the nearby forests. This second option becomes attractive for the poorer urban citizens, who realize that they can use the wealth and security of the forest to supply their housing needs by building low-cost tree cabins in the canopy.

Accra 2121 will begin simply, with a few families moving from their drowned shantytowns into the surrounding forest during a flood and resurrecting their homes out of harm's way. More will join them, including migrants from the hinterlands, and together they all will gradually learn to grow their own food and recycle organic waste within the forest in a sustainable manner. The forest's value to them will encourage Accra's new tree-citizens to protect the forest from those who would clear it—from the logging, mining, and oil companies, for instance. Currently, these industries contribute to Ghana's standing as the country with the highest deforestation rate in the world.

The vision of Accra 2121 presented here will likely be disparaged by those who are profiting from Africa's current resource boom, but for the people living in the slums and shantytowns of modern-day Accra, the idea of being able to live in a safe treehouse with your family and to secure an income from harvesting forest products before sharing them sustainably with your neighbors— this is positively utopian.

Accra 2121

□ □ □

Almaty 2121 ⌗ City of Apple Trees

Almaty is the biggest city in Kazakhstan, with a current population of nearly two million people—half of them ethnic Kazakhs, a third of them ethnic Russians, and the rest of various Asian ethnicities. Almaty is situated in the far south of the country beside the northernmost peak of the Tian Shan Mountains. The foothills nearby are held to be the geographical birthplace of the primordial apple, and Almaty is labeled *The City of Apples*—with historically famous urban stands of apple trees dotted around the city. However, according to locals, the numbers of these stands have drastically dwindled to near extinction over the past few decades.

Apple trees are not the only significant loss for Almaty. Its status as the national capital was taken away in the late 1990s. For most of the twentieth century and for all of the Soviet era, Almaty was the seat of government for the Kazakh Socialist Republic, but this status changed soon after Kazakhstan broke off from Soviet Russia, when the Kazakh parliament and government offices were transplanted north to the brand-new city of Astana.

The main reason for the transfer—though it was never officially announced as such—was to bolster Kazakh numbers and Kazakh identity over the so-called beached Russians who predominated in the northern parts of Kazakhstan. Thus, the move was an effort to deter Russian separatism within Kazakh borders and also to deter irredentism. (*Irredentism* is the name given to efforts by one nation to reclaim lost homeland from another nation.) Currently, Russian irredentism is being played out in Ukraine, Georgia, and Moldova, and it may well grow stronger later in the century with regard to Russia's aspirations in Central Asia.

In the early twenty-first century, Almaty city authorities tried to accept the transplantation of the capital with good grace and went on to rebrand Almaty as the "First City" and the "Southern Capital" as well as the "Industrial Hub of Kazakhstan." However, there is evidence that Almaty's general populace did not harbor much respect for the relocation, since it seemed to them that the Kazakh political elite were abandoning ordinary Almaty citizens to face the city's choking air, dirty streets, and occasional earthquakes all on their own.

Whatever the reason, shortly after the transfer, Almaty authorities developed the General Plan of Almaty for 2030, aiming to create an "ecologically safe, secure, and socially comfortable city." The general objective was to promote Almaty's image as a *garden city*. However, for that they've got an uphill battle to fight. Almaty is one of the top ten most polluted cities in the world, described by many as being a huge gas chamber because it occupies a valley between mountains that tends to trap noxious air. The combination of oil refineries, metal-processing plants, industrial factories, and a growing fleet of cars (both old and new) produces an infamous smog almost every day.

Much of the time, this smog grossly exceeds recommended health standards, imposing a huge health cost on Almaty's citizens, increasing their risk of lung disease and cancer, and contributing to many thousands of premature deaths per year. Added to this, numerous other environmental problems afflict Almaty: the water is contaminated by heavy metals, household garbage often piles up in the street, open spaces are being transformed into factories and warehouses, and there's a risk of radioactive dust blowing over the city from nuclear sites in other parts of the country. These factors together have also led to the demise of the stands of apple trees.

Because Kazakhstan's oil and gas sector is booming, Almaty's heavy industry is set to expand during the coming decades, so the pollution will probably become worse, pushing Almaty even further from its garden city aspirations. The citizens of Almaty will likely feel rather aggrieved that their city has been consigned to being nothing but a huge manufacturing plant, churning out goods for the Russian market and producing tax revenue destined to flow to Astana.

For decades, though, Almaty's people will probably just put up with this situation, since at least they will be suffering for the benefit of their national economy. However, the following five forces—as they unfold singularly and together in the latter decades of the twenty-first century—will turn Almaty citizens against Astana's rule.

First, the citizens of Almaty of Kazakh ethnicity will note that, despite the monologues about Kazakh identity transmitted over the airwaves from their president in Astana, it is obvious that the Kazakh political elite in the capital are very cozy with Russian companies and Russian investors and even keep their own assets stored in Russia to (a) avoid market dives in Kazakhstan and (b) evade investigation by curious locals.

Second, an inevitable string of environmental accidents is likely to be visited upon the Almaty cityscape by Russian-owned firms (such as gas plant explosions or the contamination of drinking water).

Third, there is a marked unwillingness among the politicians in Astana to push for real restitution from Russia for past eco-crimes, including seeking compensation for the radiation that is still widespread throughout Kazakhstan from Soviet-era atomic bomb tests.

Fourth, Astana seems intent on making money for itself in ways that wreck the land in the rest of the nation. For example, some politicians in Astana are set to approve the start-up of a strange new nuclear industry in which Kazakhstan agrees to import Russian nuclear waste for permanent storage in the south of the country. If the go-ahead is given, the nuclear waste will not be labeled as such, of course; it will be called something like "reserve nuclear materials" or "pre-recycled atomic fuel" in an attempt to either hide its risky nature or make it actually look like it is an asset in some way.

Fifth, members of southern Kazakhstani tribes near Almaty will begin to notice that all the highest-paid and best-positioned civil servants are from the northern Kazakh tribes. The southern tribes are likely to feel they are being ignored, and they will begin to dwell upon the idea of how much better off they would be if Kazakhstan split in two: north and south. The ethnic Russians of Almaty will also wonder whether such a split might be good for them, since the overstressed and underprivileged position of Almaty is plain for all to see. For instance, one problem Almaty residents are continually angry about today is that they are being sold overpriced, poor-quality, cancer-causing car fuels while Astana keeps cheap, good-quality fuels for itself. Almaty's public infrastructure, too—the schools and hospitals and streets and parks—is generally seen to be notably inferior to that in Astana.

At the moment, the power of the president of Kazakhstan is so total that few politicians, even those in opposition parties, want to publicly point out the staggering industrial crimes of the government and the toll they are exacting on Almaty. However, it's likely that some future president will not be so powerful and Kazakhstan will gradually become more democratic. This may open up the space for Almaty politicians to proactively call for a halt to nuclear waste imports. If this call attracts public support, these politicians are likely to see the value of pursuing a regional eco-Almaty defense against Astana and campaign for a radia-

Almaty 2121

▢ ▢ ▢

tion-free Almaty with clean air and clean streets—adorned once more with apple trees, no less.

It is sure to take decades, but before the dawn of the twenty-second century, the following policies will be enacted:

- Clean air acts will be introduced to ban the worst pollutants.
- Cars will be taxed heavily for entering central city streets. Russian-made cars will be taxed even more. Eventually, the taxes will become so onerous that people will find alternative ways of getting about.
- Bike lanes, public tramways, and Almaty-made electric eco-vehicles will provide free transportation for registered non–car owners in the city.

- As the oil and gas reserves run out, the Almaty economy will switch to light industry, services, and urban horticulture.
- The right of citizens to stands of community apple trees will be formulated as part of an Almaty city constitution. This means that stands of apple trees will have to be replanted. Investment in their well-being will be seen as a cultural benefit and a sign of Almaty's civic pride.

By 2121 AD, Almaty could be listed as one of the top ten cleanest cities of the world. Visitors from afar will come to see the City of Apple Trees standing up against the might of Russia and the petro-dictators of Astana.

Ecotopia 2121

¤ ¤ ¤

Andorra la Vella 2121 ⌘ No, No! to Nano

Andorra is a small alpine nation squashed in the Pyrenees Mountains between Spain and France. It is well regarded for its beauty, quaintness, and isolation from the rest of Europe. The capital city, Andorra la Vella, is home to just over twenty thousand people. It has no airport and no train station. This isolation, in concert with the natural resins of its alpine trees (which protect them from all manner of creatures, large and small), means that Andorra 2121 has survived the global spread of nanotechnology relatively untouched—and the locals aim to keep it that way.

Nanotechnology consists of machines and materials made at the *nano* scale, the infinitesimal scale of atoms and molecules. Nanotechnologists today promise they will soon make intelligent nanomachines to do wondrous things. Among the many claims are these: (a) nanomachines will cure the human body of incurable diseases; (b) nanomachines will smarten up our everyday dumb objects by implanting them with a proliferation of interconnected nano-sized supercomputers; and (c) nanomachines will be dispersed over land, sea, and air to clean all the pollution from the planet.

By the end of the twenty-first century, there is a slight possibility that a few of these projects will have succeeded, but nanotechnology has a dark side: providing companies and governments with powerful surveillance and weapons systems and creating new, invisible, and uncontrollable pollutants. For as much pollution and disease that it clears up, nanotech will generate as many new pollutants and diseases—and humans have virtually no experience in effectively managing these new pollutants, and no innate immunity to the new diseases. Because of their tiny size and their blundering human-programmed intelligence, nanomachines could easily escape from labs, factories, and human hosts into the environment, infecting animals and plants, killing some and disabling others. The natural world could be irrevocably damaged.

Andorra la Vella, by good fortune, manages to escape these negative impacts, which just stiffens the resolve of the city to work at preserving its nano-free status.

Antalya 2121 ¤ The Golden Orange

Antalya is city of a million people on the Mediterranean coast of Turkey. The tourist brochures promote Antalya as one of the most visited cities in the world, attracting tourists from all over Europe to its sunshine, seaside, and historical setting. Once upon a time, Antalya was ruled by the Greeks, then the Romans, and then the Ottomans, all of whom left architectural traces dotted around the city. Sometimes the reality of Antalya isn't nearly as nice as the brochures indicate, for it suffers from heavy traffic, heavy smog, and heavy, gray, lifeless architecture.

In the early twenty-first century, the city engineers are pushing for Antalya to become a *solar city*. There are many formidable barriers to this, though—political, technological, and financial—so progress is slow and faltering. In this scenario, early-twenty-second-century Antalya has become the Solar Capital of the Mediterranean, featuring solar-powered schools, solar-powered transport, solar-powered factories—solar-powered everything. The effect has been to transform Antalya 2121 into a smog-free city and to convert the gray cityscape into a gleaming golden-orange vista. In 2121, solar cells can be easily printed, pasted, or painted onto various structures in any color.

By popular vote, the theme color in Antalya has become a sunny golden orange—settled on in homage to the oranges of the same name that grow nearby and to the city's popular Golden Orange Film Festival.

Today, energy experts bemoan that one of the biggest drawbacks of solar power is its inefficiency on cloudy and overcast days, but in Antalya 2121 this drawback has been overcome in three ways:

1. *Superefficient solar cells:* These new varieties work well even in minimal light like that from a full moon. They include solar cells made from new minerals such as perovskite rather than silicon.

2. *Water batteries:* During the day, the energy from solar panels is used to pump water up to reservoirs in the nearby Taurus Mountains. When the sun goes down, this water is slowly released downhill to power hydrogenerators near the bottom.

3. *Bladeless wind turbines:* These turbines shudder just a few inches in Antalya's sea breezes, and the resulting vibrations are then converted to electricity to augment the city's nighttime energy needs.

Because this unique solar technology is manufactured using innovative processes, Antalya's expertise also provides for a thriving solar economy. Antalya 2121 has thus become the largest solar-servicing center for the Mediterranean region, along whose coastal shores live two hundred million city dwellers, all interested in buying into Antalya's solar technologies.

¤ ¤ ¤

So how might Antalya 2121 come about? The social lubricant is the sudden rise of the political process called *demarchy*. Demarchy is the selection of government by random lottery. Some people say it is a purer form of democracy because the corrupting influence of party politics and election campaigns is avoided and the composition of the government ends up being a more accurate representation of the electorate. Demarchy comes to Antalya in a fittingly unplanned, random manner. Realistically, this is the only way it could emerge, since those with political power are reluctant to give up the system that gave them that power. So what type of situation might create this new democratic landscape?

In the early twenty-first century, Turkey suffers under an authoritarian government. At the same time that Turkey's leaders pursue tight control over their citizens, they also want to show the world they are a force to be reckoned with on the global stage. Having been rebuffed by the European Union every time the country sought membership, Turkish leaders in the future decide that one of the ways Turkey can impress the world is to build a series of nuclear plants. These, they believe, will offer the double benefit of cheap electricity and the material for atomic weapons. So in the early to mid-twenty-first century, laws are made, plans are laid, and construction is begun on ten new nuclear plants along the Mediterranean coast, which are to be up and running by 2050.

As if to snub the importance of Europe, the Turkish leaders decide to build all the nuclear power plants very near Antalya, where European vacationers like to spend their summers. The Antalya provincial government at this time also invests in the project, looking forward to becoming the wealthy mega-energy capital of the Middle East. But, after a decade of planning mistakes, construction errors, corruption, and financial chaos, as well as a few scary earthquakes and fretful public protests, the half-built nuclear plants are all mothballed. The whole fiasco has plunged the province of Antalya into debt, not to mention making its electricity supply unreliable, and the provincial leaders involved in promoting the project are unceremoniously ejected from office.

Finding replacements is not easy, however. Nobody wants to be in a provincial government with no money and no energy. So it's agreed that the new leaders are to be picked via lottery, the tickets for which can be bought, one per person for one Turkish lira, on any street corner. Out of a sense of duty, or maybe a desire for power, or just the need to land a job, about ten thousand residents buy their one ticket apiece. The girl in her late teens who happens to win, with no money to run the city, isn't afraid to ask for immediate assistance from Europe to "solarize" Antalya and get the electricity flowing again. Europe is more than keen to help with various loans and technologies, since at least solar energy doesn't furnish Turkey with source material for nuclear weapons, as nuclear plants would.

By 2121, the tourists flock to Antalya again, as much in admiration of its golden-orange architecture as a yearning for its sunshine. Many of them return home with the idea of demarchy planted firmly in their minds.

Athens 2121 ⌘ The Advent of Green Trade

Athens, one of the oldest cities in the world, has a population of three million. It has been hard-hit by financial problems in recent years, and Greek master planners have experimented with various kinds of radical remedies for its economic ills: austerity, stimulus, debt default, and a possible exit from the eurozone. All the while, there have been swings in government from radical left to radical right. This is likely to go on for decades to come.

Meanwhile, the environment continues to suffer, especially with regard to *Nefos*, urban smog. Once upon a time, Nefos was a come-and-go visitor, appearing maybe one day in four, but with changing climate patterns and lax regulations, the smog has become ever present, ravaging the lungs of young and old, spoiling the view, and eroding ancient monuments. Even now, when Nefos is in town, six times as many people die compared to clean-air days. Nobody wants to come to Athens anymore—it's impossible to even see the Parthenon.

As the decades of the twenty-first century pass, the trade unions find themselves gaining power and influence. When a left-wing government is in power, the laws related to collective bargaining become stronger. When a right-wing government is in power, the membership base of unions swells as part of a backlash against cuts in public funding. With the left wing and the right wing fighting over taxes and pensions, it is the unions in Athens 2121 that have become the powerhouse behind environmental change. The bus drivers' union, for instance, forces through a citywide Green bus program to allay the daily onset of Nefos, then goes on to claim that Athens's roadways should be for the buses alone, forcing private cars off the road.

Factory workers strike against hazardous pollution in their workplace and call for cleaner manufacturing. The trend continues when the maritime workers' union refuses to load or unload goods that are not eco-friendly or that arrive on ships that are not eco-friendly. As in other key ports around the Mediterranean, the unions in Athens—like their new best friend, Antalya 2121—push for safe, Green trade around the world. Nefos slowly disappears from the Athens sky, and both the tourist industry and the health of Athenians perk up again.

Bastia 2121 ⋈ A War against Eco-destruction

Bastia is the major port of the French island of Corsica. In 1972, an Italian company dumped toxic waste off the Corsican coast, creating a grotesque red mud around the island and causing dead porpoises to wash up on the shores for weeks. Corsicans found the French government to be terribly lax in its response, for it neither lodged a complaint against Italy in any way nor honored pledges to rehabilitate the environment.

At the time, France and Italy were negotiating trade agreements, and the leaders didn't want eco-quarrels to get in the way. This negligence seems to continue into the present. Italian tankers leak oil near Corsica, and Italian cruise ships sail blithely through vulnerable Corsican marine zones. There are suspicions, too, that radioactive and industrial waste is being dumped near Corsica by the Italian energy ministry. The Corsicans have taken legal action, but Italy still acts as if it owns the sea around Corsica.

Perhaps Emperor Napoleon's legacy has had some influence on this state of affairs. Although he was a native of Corsica, having been born there in 1769, he spoke only condescendingly about the island and never cared to defend it. In this scenario for Bastia 2121, the Corsica of the future has given rise to another Napoleon—Napoleon of Bastia. Born in 2069, three hundred years after Napoleon I, he rises through the ranks of the island's government to become governor of Corsica. Unlike his namesake, though, he has the island's well-being at the front of his mind. The number of environmental laws approved for the island skyrockets. For instance, he resurrects a sixteenth-century law mandating that all Corsicans plant four trees a year (a fig, a mulberry, an olive, and a chestnut). To top it off, in 2120, he takes a strong stance against an Italian garbage-dumping vessel heading to Corsica, and he sends out fighter planes to fire warning shots at them. The world is astonished to see military action taken to preserve the environment rather than to destroy it. The next year, in 2121, Napoleon of Bastia is voted president of France.

Beijing 2121 ¤ City of Gold, Part I

The Chinese have a special fondness for gold, regarding it as *pure, lucky, noble,* and *glorious,* and as being more trustworthy and incorruptible than just about anything else in the universe.

The Chinese government shares this feeling, and it has set about getting its hands on as much gold as it can. According to some, gold hoarding is part of China's attempt to *de-Americanize* the world by challenging the trading power of the mighty US dollar. Currently, the exact total of government-owned gold in China is a mystery. However, someday in the mid-twenty-first century, Beijing will declare its gold holdings in the most conspicuous way: the Gold City of Beijing, a new commercial center where the buildings are fabricated from gold, will display for the world to see the glorious financial power of China.

At first, the Gold City is planned to be glamorous and showy. However, things may change a decade or so later, as China suffers from myriad large-scale environmental crises:

1. The air becomes so polluted that it's no longer possible to hide the fact that it kills tens of millions of Chinese citizens per year. Even now, just by walking to school and home again, many urban Chinese children suffer the effects of smoking the equivalent of two packs of cigarettes a day. By the middle of the twenty-first century, with continued industrial expansion, it may be much worse.

2. A bunch of nuclear plants near China's coastal cities are hit by violent natural disasters: cyclones, earthquakes, tsunamis, and storm surges, provoking multiple meltdowns, explosions, and the release of radioactive clouds into the atmosphere. As if that isn't enough, soon after, hordes of rats and swarms of locusts come and go within the battered nuclear plants with gay abandon, and when they leave, they are contaminated with radioactivity, which they spread far and wide among urban and rural settings. Because of the contaminated clouds and the radioactive rats and locusts, entire cities have to be evacuated and abandoned.

3. Record-breaking monsoon rains during a single season cause the worst series of floods in Chinese history along the Yangtze, Yellow, and Pearl Rivers. These floods force the evacuation and eventual abandonment of China's river cities, a process that affects many millions of people.

4. It is announced that the panda has become extinct in the wild, and the remaining zoo specimens die out because they are unable or unwilling to reproduce.

These calamities could very well play out in quick succession sometime in the mid-twenty-first century in the years set to become known as the Dirty Decade in China. Together, these disasters will force the Chinese populace into a massive reevaluation of industrialization and the way the country is governed. By the end of the Dirty Decade, confidence in the Communist Party will have shrunk beyond recovery and massive public dissent will become widespread. Because people will demand that their voices be heard in the actual management of the ongoing environmental crises, calls for democracy and human rights will become louder and more strident than ever before. The level of protest will be just too raucous for the authorities to contain or shut down, and they'll be forced to concede to democratic reforms.

◻ ◻ ◻

In order to survive, the Gold City must adapt its image to these changing circumstances. Originally, the Gold City was designed to be a demonstration of Chinese success—of its glory, wealth, and power. But as the years roll by and the building plan is put into effect, these sentiments do not reflect the changing environmental concerns of the Chinese public, and the project has to be redeemed somehow. The answer is this: the Eco-gold City.

Currently, the world's gold is pulled from the ground in environmentally and socially suspect ways, usually involving forest clearing, labor abuse, the destruction and removal of topsoil, the disruption of small communities, and the contamination of waterways both on the surface and underground. Chinese gold-mining companies are as bad as any with respect to these issues, whether operating at home or abroad. Chinese gold mines in Tibet and Africa, for instance, have caused landslides, land clearing, and heavy-metal contamination, and resulted in various social ills ranging from the trafficking of sex workers to the corruption of local governments. Chinese mining companies are usually among the most strident in rejecting any form of investigation into their practices or regulation of their activities. Meanwhile, they often make no discernible contribution to the economic improvement of the areas where they mine.

Gold mining is sometimes held to be one of those industries that are inherently dirty, and no amount of regulation or Greening will change that. However, if the business model is made to change, and if only nontoxic gold-mining techniques are used, and if the industry is policed according to international law and fair labor agreements, and if labor unions are allowed to organize, and if wilderness protection is undertaken (and supervised by top managers, local authorities, and national and international agencies, as well as by the local communities, nongovernmental organizations, gold traders, and consumers, all collaborating in the effort), then it is possible for gold mining to become far better than it is today.

Beijing 2121 represents this to some extent. Here, each brick is an eco-gold brick, produced by a gold industry universally adjusted by having adopted three fundamental policies:

1. *The No Stench Policy:* Gold mining should not change the natural scent of the air, thus the predominant flora must be preserved, noxious chemicals must be abandoned, lakes and fisheries must be conserved, and mining sites must be restored.

2. *The Random Watchdog Policy:* Randomly chosen community members are enlisted to supervise the operations and finances of gold miners (with guidance from randomly chosen international scientists, lawyers, and accountants).

3. *The Common Heritage of Mankind Policy:* Gold mined from publicly owned land should be declared the Common Heritage of Mankind, whereby gold can be rented year by year and shifted around the world, but it cannot be bought and sold. The ongoing revenue from gold rental can then be used for environmental benefits in the country of origin.

Bethlehem 2121 ¤ City of Gold, Part II

Bethlehem is a Golden City of an entirely different kind than Beijing—it is golden in terms of spirituality. Bethlehem is held to be the sacred birthplace of such divine figures as Lachama, the Canaanite God; David, the King of Israel; and Jesus Christ, the son of the Christian God.

Control of Bethlehem has shifted throughout its three-thousand-year history. Once the domain of the Canaanites, it later was ruled by Judea, Rome, Persia, and an Arab caliphate. Next it became part of Egypt, then it was claimed by Christian crusaders, followed by the Ottomans, and in the twentieth century it was governed by Britain, Jordan, and Israel. Since the 1990s, Bethlehem has been part of Palestine.

Bethlehem's potential for future ecotopian status is profoundly linked to its increasing independence from Israel. Compared to many other parts of Palestine, Bethlehem is less reliant on Israel for water, trade, or security. With this comparative independence, Bethlehem utilizes its golden status in spirituality to appeal to a broad base of international tourists, from which industry most of its citizens can derive a healthy living. To preserve this economy, Bethlehem 2121 has banished industrial zones as unsuitable for the city's character. Instead, only small craft and cottage industries such as the production of olive oil and date products are approved.

Bethlehem 2121 stands in stark contrast to the rest of Palestine, which has become a hodgepodge collection of industrial parks set up by foreign firms to take advantage of the cheap labor and lax regulations there. For some Palestinians, these industrial estates are considered *zones of prosperity* because the new industries offer jobs to locals. However, all too often they end up being *zones of environmental injustice*, populated by dangerous and polluting industries. In contrast, Bethlehem has clean air and clean water, and its people are healthier and happier as a result. And as long as the economy remains based on tourism, handicrafts, and small-scale agriculture, there's not much pressure on infrastructure from population influx.

Birmingham 2121 ¤ The Green Heart of England

Birmingham is England's second-largest city. During the Victorian period of the nineteenth century, it was famous worldwide for being Britain's manufacturing heartland. Some nicknames coined for the city at that time include "The Workshop of the World" and "The City of a Thousand Trades." Nowadays, the manufacturing sector of the city has dwindled and been replaced by a service economy. Even the iconic Mini Cooper automobile, once made in Birmingham, is now produced abroad. The "heartland" tag applies only to the city's geographical location, for it is approximately in the middle of England.

Before it grew into an enormous manufacturing center, Birmingham was just a small hamlet within the medieval Forest of Arden. During the Middle Ages, the Arden woodland contained oak trees intermingled with chestnut trees, and many birch and linden trees as well. These are long-lived species that host all kinds of animals—large and small, rare and common.

For three thousand years, the density of the Forest of Arden made it nearly impossible for succeeding civilizations to settle within or build roadways though this woodland. The Celts, the Romans, the Anglo-Saxons, and the Vikings all left it intact. There were clearings—*leighs*, they were called—among the oaks and birches, and many tiny villages were located in them. For millennia, these villages utilized the resources of the forest sustainably at a small scale. By long tradition, much of the Arden during this time was common land: open equally to all village families and guaranteeing them certain rights to graze animals, to collect wood and char, and to hunt and forage.

After the Norman invasion of England, toward the late eleventh century, a slow, five-hundred-year colonization of the forest took place. The new king, William the Conqueror, was not really of an egalitarian persuasion, and he stratified the medieval society of middle England by granting power and resources only to those most acquiescent to his goals. This included the very Catholic Arden family, who were given the right to control vast tracts of the Arden.

Slowly, as the centuries passed, most of the forest was enclosed within fences, then privatized and converted into farmlands—mainly for the lucrative wool trade. Sheep and wool made England a wealthy nation by the end of the Middle Ages, but they also led to the decimation of its largest forest.

More than five hundred years after the Norman invasion, in the decades of the late sixteenth century, the Arden family had grown to be very large and wealthy in the English Midlands. However, because they were Catholic, they lacked significant political influence, since by this time the English monarchy had changed to being Protestant, while Catholics were actively marginalized.

One of the members of the Arden family at this time was Mary Arden, the mother of William Shakespeare, and he too was born in the area that would have been within the boundaries of the ancient forest. However, by the time Shakespeare achieved fame, there was virtually nothing left of the forest—save for some isolated stands of woodland.

This wouldn't stop Shakespeare from waxing lyrical about the Arden on numerous occasions, though. For example, here's a passage from the comedy *As You Like it*, wherein an overthrown duke has exiled himself to the Forest of Arden to escape the clutches of the new, warmongering, tyrannical duke, who also happens to be his brother:

Sweet are the uses of adversity,
Which, like the toad, ugly and venomous,
Wears yet a precious jewel in his head;
And this our life exempt from public haunt
Finds tongues in trees, books in the running brooks,
Sermons in stones and good in everything.

If that's not expressive enough, later on another character, Orlando, is found singing about the Arden to his faraway love:

Under the greenwood tree
Who loves to lie with me,
And turn his merry note
Unto the sweet bird's throat,
Come hither, come hither, and come hither:
Here shall he see
No enemy
But winter and rough weather.

Shakespeare wrote these lines about the Arden being a haven in 1601, but—just four years later in a village not far from Birmingham—the same area served as the birthplace of England's most illustrious revolutionaries, the conspirators behind the Gunpowder Plot. The plotters, who included in their ranks the now notorious Guy Fawkes, aimed to assassinate King James, the Protestant head of England, whom the plotters regarded as an

Birmingham 2121

oppressor of their Catholic faith. Their plan was to travel from Birmingham to London on horseback on the night of November 5, 1605, then blow up the entire House of Lords from underneath, thereby killing King James and all his Protestant ministers. The plot was foiled when Guy Fawkes was caught guarding more than twenty barrels of gunpowder in the basement. The rest of the Gunpowder plotters rode back to the Arden and hid out in a small settlement near Birmingham before being hunted down and killed by a militia loyal to the king.

Shakespeare's father was actually good friends with the leader of the Gunpowder plotters. However, nobody has yet found out Shakespeare's true religious convictions and whether he sympathized with the Gunpowder plotters or not. But in both fiction and history, the woodlands around Birmingham have served as a green haven from corrupt authority.

¤ ¤ ¤

Four hundred years on, in the twenty-first century, the story of the Gunpowder Plot survives as Guy Fawkes Night, when the English come together on the evening of November 5 to light bonfires and set off fireworks. Unlike the Guy Fawkes story, though, the Forest of Arden has not survived so well. Today, there are just a few tiny remnant groves of the ancient forest left. Some of these have one-thousand-year-old oak trees within them and linden twice that age. These small, ancient stands serve as the last irreplaceable habitats for Middle England's rarest species, such as the turtledove, whose beautiful call has inspired many artists over the centuries, and the natterjack toad, so stereotypically warty that it needs more than Shakespearean romance to render it handsome.

Looking hopefully into the future, we might believe that these ancient remnants of the Arden, and the wildlife they are home to, will soon be appreciated as havens from city life and therefore worthy of conservation. Perhaps the residents of Birmingham will see to it that each individual isolated stand of ancient oak on the city's northern and southern boundaries is cultivated to connect and converge with the others. Slowly, green corridors could multiply throughout the entire city, allowing the forest to slowly regrow and become sustainable again. This is the scenario for Birmingham 2121, where the ancient Forest of Arden reinhabits almost every neighborhood of Birmingham.

To look romantically upon the past is not unusual in Birmingham. The city has many monuments and museums devoted to its nineteenth-century history, when it was reckoned to be one of the greatest cities of the industrial world. Sometimes it seems there are more museums of industry in the city than working factories. But why must Birmingham's nostalgia be confined to the Industrial Age? Why cannot Birmingham's cultural keepers cast their net wider, be more imaginative, and dream of Greener times?

The idea of resurrecting the ancient forests of England's Midlands is not entirely new. In the 1980s, the English Countryside Commission designed the new National Forest in the approximate territory of the extinct Arden Forest, except it excluded large cities like Birmingham. The commission stated the new forest would achieve multiple goals:

- It would create habitat for thousands of species, helping to nurture English biodiversity.
- It would sequester greenhouse gases from the atmosphere in wood and biomass, thereby helping to lessen the impact of climate change.
- It would reduce atmospheric pollution, filtering out smog particles.
- It would significantly help in flood control and the purification of water.
- It would nurture the precious Midlands soil; indeed, in places, it would save it, and create more.
- It would be a place of beauty, calm, and recreation, providing Birmingham residents and all of England with outdoor activities and a link to their environment and heritage.

While conservationists loved the idea, for various political and economic reasons the plan was never enacted by any level of government.

However, perhaps private enterprise can step in where government has feared to tread. The Heart of England Forest Ltd. is a woodlands charity set up by the late businessman-turned-philanthropist Felix Dennis. Dennis was a maverick in publishing circles. One time in the 1970s, he was taken to court for letting high school students edit an issue of his magazine, which ended up including a sexually suggestive parody of Rupert Bear, a popular cartoon character. Dennis was charged with "corrupting minors." In order to raise money for his legal defense, John Lennon offered to record a song with him. Eventually, the court case was decided in Dennis's favor, and he went on to greater fortune by publishing a variety of pop magazines.

Some of this fortune is now in the process of being utilized by Heart of England Forest Ltd. to buy thirty thousand acres of land just north of Birmingham. This land, entirely open to the public, is specifically for the purpose of planting new forest and creating new habitats for the region's wildlife.

If the plan succeeds, and if it achieves broad popularity, then it is possible that Birmingham's leadership will offer support one day in the future for the forest to grow tree by tree, stand by stand, into their city. Among the leighs, through the woods, perhaps then we will see the diminutive Mini Cooper again, made in Birmingham once more and resurrected in electric-motor mode, its battery charged by the common household wind turbine.

Ecotopia 2121

¤ ¤ ¤

Bristol 2121 ¤ The Eco-bridge

Bristol, like Birmingham, claims to have been a core city of the nineteenth-century Industrial Revolution. It was the hometown of the famous engineer Isambard Brunel, builder of the world's greatest bridges, canals, and railways.

Bristol today claims to be a Green Capital of Europe, having officially been awarded the title in 2015. This honor was met with varying degrees of skepticism by Bristol's residents, since many note how bad the traffic is and how badly polluted some of the city's waterways are.

In Bristol 2121, the engineering vision of Brunel is combined with a massive renewable energy project. A tidal bridge between Bristol and the Welsh city of Cardiff has been built across the Bristol Channel. The channel has one of the strongest daily tidal movements in the world, and, as the water flows in and out every day, the energy is harvested to power much of southern England and southern Wales.

As well as this, a long, slender city has been built upon the bridge—a mixed zone for residence, retail, entertainment, education, small industry, and health care—everything a normal city needs. What the bridge city does not have are huge roadways, since no one needs a car. As a mixed zone, people can live and work and shop and play and go to school, all within walking distance. The bridge city is populated by skinny architecture. Skinny buildings consume less space and require less servicing. Also, fewer materials are used in their construction, so they are more eco-friendly and cost-conscious. Many of these skinny houses also have microgardens rather than huge, sprawling backyard lawns. This means their water needs are less than half those of the average suburban home. Because each building is attached to the one next door, they are twice as easy to heat or cool as conventional detached houses in Bristol.

Budapest 2121 ¤ Hungary's New Gardens

The scene shown here depicts the Hungarian parliament on the banks of the Danube River on the *last* day of 2121. Surrounding the main building are plots of organic vegetables grown by various members of the parliament, who share and barter the produce among themselves and with people from around the city.

So how did we get to this ecotopia of garden delights? On the *first* day of January 2121, the president of Hungary stood on a podium in front of the nation lamenting that "weird hippie lefty factions" had invaded the parliament and were undermining the culture of the nation and the security of the state. He begged the parliament to invest more powers in his presidential office to circumvent this degradation of the nation. To bolster his argument he related how one MP from the Hungarian Green Party grew vegetables in his parliamentary office in a "stinking pot of his own excrement!" The speech was fervently discussed in the media and among ordinary folk by word of mouth. Nationalists retold the story with anti-Greenie invective and jumped to support the president and give him more powers. However, by the time the parliament opened for assembly in late January 2121, something strange had happened. Many moderate politicians had begun their own in-office gardens. They weren't brave enough to use their own bodily waste as soil, but they were smart enough and well-organized enough to procure organic compost from local suppliers or from their own countryside lodges. They then broadcast images of their new office gardens.

Overall, the public feedback about these gardens was very positive. The attention encouraged urbanites all over Hungary to take up the cause as well. A new fashion spread across the country like wildfire: office workers grew vegetables on their balconies to eat for lunches, children grew their dinner vegetables in school gardens, and pensioners got together to grow winter soup veggies in abandoned lots around the city. It was an eminently practical trend, drawing on long-lost traditional methods of gardening to augment the family diet. The practice also sent a clear and blunt statement opposing the president's grab for more power. By the last day of 2121, the parliamentary vote to cede authority to the president was defeated by the MPs, who were noisily munching on crisp green peppers.

Burlington 2121 ¤ Anarchy in the Yew Bay

Burlington is a Vermont city of fifty thousand people situated on a yew-lined bay of beautiful Lake Champlain. As well as being the birthplace of Bernie Sanders's political career—he served as mayor in the 1980s—Burlington is also the birthplace of Ben and Jerry's, an ice-cream company celebrated across the continent for its creative menu, community projects, and organic, eco-friendly ice cream.

Unbeknownst to most Burlingtonians, the city also hosted another celebrated resident, at least in the environmental community: Murray Bookchin, a theorist of eco-anarchism. Like other brands of anarchism, eco-anarchism has a strong disdain for authority flowing through it. Bookchin taught that all beings of the world, human and nonhuman, have the right to be free from authority. This means eradicating human domination of nonhumans as well as human domination of other humans.

The big antiauthoritarian endeavors most eco-anarchists aim to peacefully (without the use of arms) rid the world of are: (a) corporate power, (b) central government, and (c) national armies. Only then, the eco-anarchists say, will both humanity and nature be at peace and free.

This freedom would allow communities to make their own decisions, which, Bookchin suggested, would generally favor the long-term preservation of the natural environment rather than its destruction for short-term gains of power. It would also reassert the local quirky differences of a place over the rampant unformity imposed by corporations and government.

As the twenty-first century proceeds, increasing numbers of environmentalists from around the world visit Burlington and end up converting Bookchin's old house into a mini-university for the study of eco-anarchy. Over time they initiate several urban projects, the most popular being to convince Burlington that a city government isn't really necessary. By the end of the century, the city council votes to disband itself. The tasks that the council once performed are dispersed to groups of enthusiastic people who enjoy mastering them and teaching others about them. Other tasks are distributed to various small businesses vetted at public assemblies for their trustworthiness. The world may continue on its merry way with large corporations and armies and governments, but Burlington follows its own quirky, anarchic path to ecotopia.

Cape Town 2121 ⌧ The Fission City

As South Africa ponders its energy problems in the mid-twenty-first century, offers of assistance come thick and fast from international friends Russia and China. Like South Africa, they are members of the BRICS group of top emerging economies, and they offer to give South Africa free uranium that they've mined from nearby Namibia. Russia and China also propose to build cheap, "at-cost" nuclear plants for South Africa in exchange for some favorable investment help in Africa. If Russia or China can gain a nuclear beachhead here, they could likely make future sales throughout the entire continent.

The Russians and Chinese probably zeroed in on South Africa because it was the first (and so far only) African nation to build commercial nuclear plants. The South African government is also currently testing the waters to see if it is possible to expand its nuclear sector.

In this scenario, Russia's offer involves building the nuclear plant in Namibia, then transferring the electricity into South Africa. The Chinese proposal would ship processed uranium from Namibia to a nuclear plant that is set up right next to a major South African city. This second proposal would save money on infrastructure cost, and Cape Town is touted as the preferred host site.

Because the Chinese proposal is cheaper, this is the option South Africa chooses, and in the late twenty-first century a huge, sprawling nuclear complex is built in the suburbs of Cape Town. Both South Africa and China herald the project as transforming Cape Town into a Green Energy City, since nuclear power plants produce no carbon dioxide and they will enable Cape Town to phase out their coal-burning plants.

Many South Africans are wary about nuclear energy being classed as Green, but the president is convinced that the nation has no time to "experiment with untested renewable energies." The danger of global warming, he says, is too imminent for that: "We need a fully developed carbonless energy source, and we need it now!"

□ □ □

Nuclear energy is not going to rescue the world from global climate change. It is true that nuclear power stations produce minimal carbon dioxide during their operational phase—the phase when uranium rods are sitting in a reactor working to produce electricity. Yet to get uranium into a reactor in the first place requires that great quantities of carbon dioxide be emitted through the long and complicated processes of uranium mining, uranium enrichment, and the construction of suitable power plants. Also, at the back end of the nuclear cycle, when spent uranium rods need to be disposed of and when plants have to be decommissioned, dismantled, and decontaminated, fossil fuels are likewise used in massive amounts.

It is true that if you add up all the greenhouse gases that are produced to commission and decommission a medium-sized commercial nuclear power plant, the amount still comes out lower than the emissions from a medium-sized coal plant. However, if we compare nuclear plants to similar-sized renewable energy projects—say, solar or wind power—then nuclear power is five times more polluting.

Alas, none of this may stop the South African and Chinese governments from pushing for Nuclear New Build together later this century. In order to reduce global carbon dioxide emissions enough to avert massive climate change on a global scale by 2121, a Nuclear New Build strategy would call for at least two thousand new nuclear reactors to be built worldwide within the next decade. This is a huge number, considering that there are only some four hundred reactors operating today, and they were built over a sixty-year period. Each power plant takes a decade to get going, from the breaking of new ground to the production of electricity. If there is public reaction against the plant, a certainty in all democratic nations, you can add another ten years for consultation and plant redesign.

The only nation in the world at the present time with the impetus to invest in large-scale Nuclear New Build is China. But China also has very dodgy safety standards and usually rides roughshod over both public sentiment and risk concerns. A crazy, full-speed-ahead Nuclear New Build by China is bound to result in a large-scale nuclear accident sometime. It might seem cheap for South Africa to plop a Chinese nuclear power plant next to Cape Town, but it certainly will not be safe. Nor, probably, will Cape Town residents be as quiescent as China's citizens.

But even if everything is a success—even if accidents don't happen, construction isn't delayed, public protests do not erupt, and the cost overruns can be managed—a massive Nuclear New Build

Cape Town 2121

❏ ❏ ❏

program will still create environmental problems by producing huge amounts of nuclear waste. In the future, China may promise to "repatriate" this waste for a fee, but this will still mean that the waste will be stored in Cape Town for decades, contaminating water supplies and inviting nuclear terrorist attacks, before finally the nuclear waste is taken by China and probably dumped in an abandoned Chinese-owned uranium mine in Namibia.

So how could a huge, sprawling, dangerous, and polluting nuclear plant built in the suburbs possibly create an ecotopia out of Cape Town? The answer is obvious. Do not turn it on! Sure, billions of dollars will have already been sunk into the project by then. And, sure, hundreds of millions more will have to be paid to China for "breach of contract." However, these costs are but a fraction of the cost of managing all the risks, dangers, and pollution that are likely to occur over the lifetime of the plant if it is turned on.

Ecotopia 2121

¤ ¤ ¤

Chicago 2121 ¤ The Fusion City

Now I am become Death, the destroyer of worlds.

Robert Oppenheimer, "father of the atomic bomb," used this line from Hindu scripture after watching hundreds of tons of New Mexico desert being blown skyward in July 1945 as a result of the world's first nuclear bomb test. The first atom bomb experiments actually started in Chicago three years earlier, though. It was in a racquet court under Chicago University's sports stadium that scientists were piling up uranium rods to force a nuclear chain reaction—right within a city of three million people. The pile had no shielding, so an explosion could have sent radioactive wind over the entire city. Luckily, that didn't happen.

In Chicago 2121, the Department of Energy again thrusts Chicago into an epic atomic experiment, building within the city the world's first nuclear *fusion* plant. Conventional nuclear plants have *fission* reactors; they split atoms apart to make energy. A *fusion* reactor, on the other hand, pushes atoms together to create energy. For twenty years or more, the DoE has been telling Chicagoans that fusion plants are much better than fission plants; they are safe and produce no radioactive waste. Nor can fusion reactors be used to make atomic bombs, unlike conventional fission plants.

While Chicago vacillates in its support, a series of politicians, scientists, and industrialists working together eventually succeed in locating a fusion plant in Chicago. They attract many billions of federal dollars to fund the process. Despite that, many Chicagoans who were avidly following the nuclear decisions going on in Cape Town ask, *If Cape Town can veto the startup of a fully built nuclear plant, then why can't Chicago veto an unfinished nuclear plant?*

During the decades the plant takes to build, Chicagoans learn many things about fusion. In fact, fusion does produce radioactive waste, just not as much and not as long-lived as that produced by a fission nuclear plant. Yet the waste is still dangerous and very expensive to manage. They also learn that the core of the reactor could slowly degrade the structures surrounding it, leading to its eventual collapse.

With Oppenheimer's lamentation never far from their minds, and as they watched Cape Town turn away from nuclear energy at the last moment, Chicago's citizens—first by means of protest, then by petition, and finally by referendum—eventually force the abandonment of the project at the eleventh hour.

One gem of wisdom usually handed down to children from their elders is the story of seeds. Either at home or at school, on a farm or in a garden, just about all kids learn about how the trees towering above them all grew from tiny seeds that could be held in the palm of their hand.

In an industrial setting like Chihuahua City, the life cycle of plants is often of little concern, as citizens seek to preserve their own cycle of human life. Despite this, though, in just about every industrialized zone around the world there are what we may call *forest dreamers*: those who feel the hope, inspiration, and possibility held within little seeds, that they might one day become a magnificent forest. Many nations have had forest dreamers of legendary status: Johnny Appleseed in the United States, Elzéard Bouffier in France, Jadav Payeng in India. These are individuals who purposefully, over many years, all by themselves, sow entire forests or parklands or orchards along the sidewalks or pathways or in abandoned areas through which they happen to amble.

¤ ¤ ¤

Chihuahua City today has a population of one million people. Its economy depends upon its proximity to the United States border and upon the industrial practice of *maquilló*. Unfinished parts arrive in Chihuahua from the United States and are then assembled using cheap labor in unsafe factories before being sent back again to be sold on the US market. Some well-known companies using this business model include Ford Motor Company, Honeywell, and Hallmark.

Chihuahua's other main industry is mining. More and more mines—mainly copper and silver—are popping up on the desert outskirts of the city. The mines are increasingly the subject of environmental concern, especially with regard to soil erosion and the release of toxic heavy metals. The mining companies, such as Rio Tinto, are legally required to teach mine safety to residents, mainly rural indigenous people, near the mine sites. However, the companies usually use this as a public relations opportunity to promote the mines. For instance, they tell the residents something along the following lines:

You can carry on working to harvest fibers from the agave plants, getting pricked by their spiny shin-digger leaves every day, suffering itches and rashes. Or you can give up this old-fashioned dangerous occupation and come enjoy high wages in a safe mine.

The reality is quite different. Mercury from the mines ends up in the water, and dust contaminates the workers' lungs. Environmentalists and unionists campaign to clean up or close down the worst mines, but, alas, the mines expand across the Chihuahua landscape. In the future, many locals begin to think that a temporary rash from an agave plant is infinitely more tolerable than lifelong lung disease or cancer.

¤ ¤ ¤

Within this environment of toxic mines on the outskirts of the city and unsafe factories in the inner city, it's still valuable to fantasize about a Greener future. Our Chihuahua fantasy starts sometime in the mid-twenty-first century with a five-year-old schoolgirl, Flor, as she walks home with a bunch of seeds gathered during her morning lessons. As she walks past an abandoned canal, around the corner from a quiet factory, across the dusty, rocky field

near her home, she stabs a hole in the sandy soil, places a seed, and pours a few drops of water from a drinking bottle on it. The next day she checks on the places she has sown, but nothing has happened. And the next day, too. And one more day. Nothing.

But the walk to school and the walk back home again are boring and long, and there's little else to do, so for a week or more Flor keeps sowing seeds and then checking on them later. One afternoon, in a spot near a waterlogged hole, maybe a month or so after she began, a small green shoot with a few leaves has risen out of the soil.

As soon as she gets to her class the next day, she asks her teacher for more seeds. And every day after that as well. Sometimes her teacher obliges; sometimes she cannot.

Flor didn't know what sort of seeds she was sowing. They were all different: some small, some large, some smooth, some grooved, some even had wings. The teacher didn't know what Flor was doing with the seeds—in fact, both teacher and student thought the other a little bit weird, so they never discussed the matter at any length.

Sometimes Flor's new shoots withered and died, but sometimes they grew green and woody and sturdy. One grew into a squash. Flor liked them

Chihuahua City 2121

¤ ¤ ¤

all, and she was determined to continue, imagining that by the time she left school forever—an eternity into the future for her—a mighty forest might line her long route.

It didn't quite work out that way, but enough of Flor's seeds germinated and grew to keep her enthused while her teacher kept donating seeds. Over the next few years, Flor's friends and families heard about her efforts, and a few times she had the chance to show them, while walking around her neighborhood. Some thought her endeavor inspirational. Others thought it was an eccentric waste of effort, since sooner or later the plants would be mowed down by the city.

But by Flor's last year in high school, many of her plants were still standing and some were taller than she was. By then she had learned what kind of trees they were—mostly desert willows, as it turned out.

◻ ◻ ◻

In later years, moving into the new century, Flor's fascination with trees did not wane. She was always trying to enlist her friends and neighbors to help with weekend sowing and planting around their community. The local newspapers chanced upon their activities, and started reporting on them as "Greening the City." Over time, all across Chihuahua City, readers began offering their support by donating seeds or helping to spread them, or by pointing out neglected sites that could do with some Greening. Despite numerous setbacks—such as parking lots and new factories being built on certain plots after years of gardening there—Flor's part of Chihuahua City, on the whole, grew gradually more blossomy and botanic.

The emerging urban forest of desert willows within Chihuahua delivered a host of benefits to the city that Flor didn't initially realize:

- They helped shelter neighborhoods and communities from the adverse effects of wind and sun, keeping them cooler and more humid.
- They helped mitigate all kinds of pollution; air pollution waned somewhat, as did noise and visual pollution.
- They helped control erosion in certain places where the soil and subsoil had been made fragile by development and overuse.

At one point, Flor also started cultivating plots of agave plants. She took a fancy to one particular species, *Agave lechuguilla,* "the little lettuce," which grows only in the arid soils of Chihuahua and in no other place in the world. Despite its nickname and attractive appearance, the little lettuce is a hardy, tough little plant with rigid, saw-toothed leaves that can easily pierce the toughest of outdoor wear. With the right skill, the fibers of the little lettuce can be turned into ropes, mats, and brushes. Native Chihuahuans contend that these products can last for decades, so tough are the fibers.

Occasionally, Flor would venture off into the outskirts of Chihuahua City to see if she could find out more about the little lettuce from local people. While the plant is quite common in the desert, it never grows near a mine. Some said the plant is sensitive to pollutants, while others thought it is because the little lettuce knows where the Earth is dead and the people are bad.

By the beginning of the twenty-second century, the mines around Chihuahua had been bled dry of commercial metals, and the mining operations all rather quickly shut up shop, one after the other, within a few years. For many locals, this was not something to be worried about. The land would finally be left in peace. By now, Flor had grown to old age, yet in 2121 she was still ready to begin a new project. Under her direction, the indigenous craft of agave cultivation started up once more, to replace the mining industry. She was determined to develop a nature-based cottage-style industry so that her grandchildren would not ever have to work in a *maquilló* factory. She was such a well-known figure in Chihuahua City by this time— although she was still regarded as eccentric—that she acquired help from all sections of society: natural history clubs, the local citizenry, and some factory managers, too. Working together, by the end of the year, Chihuahua's first agave-roofed homes, made from the fibers of the little lettuce, were made available to families in different parts of the city.

The forest that Flor had dreamed of when she was five years old ended up taking an entire life-time to mature, but eventually it enveloped entire sections of the city. Her seedlings not only managed to survive the city; they ended up becoming an integral part of it.

Chihuahua City 2121

¤ ¤ ¤

Como 2121 ⌗ City of the Lake Hills

Como is a city of ninety thousand people located on the scenic shores of Lake Como in the northernmost province of Italy. It's a very old city. Around the first century BC it became subject to rule by Rome. At the time, the town center was situated in the nearby hills, but it was moved to its current location by order of Julius Caesar, who had the swamp near the south end of the lake drained before ordering the construction of a walled city in the typical Roman grid pattern.

Caesar's city walls are now mere ruins, but the town of Como is quite a delight—nestled between a lake and the Italian Alps. However, this splendor belies the toxic nature of the lake water. Although a clear azure blue, Lake Como is actually an unhealthy microbe-infested danger zone, the water unsafe for drinking or swimming. Unwary visitors to the lake can suffer a spate of health problems after any contact with its water. The provincial government has blamed this sad state of affairs on unrestricted second-home housing developments, especially for the nouveau riche wanting to show off. These luxury residences pump their untreated sewage into the lake.

A favorite pastime for the new Como residents is to cruise around the lake on lavish yachts. However, in 2121, on one early evening sojourn, a party yacht runs into a floating log, is punctured, and capsizes. Nobody drowns—they're only yards from the shore—but one partier, a world-famous superstar model, suffers a chronic facial rash that never entirely fades. Forced out of the modeling business, she retires to team up with environmentalists and press for ecological change. For Como 2121, this means reversing a two-thousand-year-old planning blunder made by Julius Caesar. The swampland is redeveloped to act as a natural filter for the city's sewage, and the town is resurrected as it had been before Roman rule, on the alpine hills nearby.

Como 2121

¤ ¤ ¤

Dawei City 2121 ⌖ Eco-power to the People

Dawei is a city on the southern coast of Burma. The true size of the city, with some 150,000 residents, is hidden beneath the coconut and betelnut palms that give the city a tropical forest ambiance. The city's residents are predominantly of Dawei ethnicity, making it different from other cities in Burma, which are dominated by ethnic Burmese.

The Dawei settled here in the eighteenth century to avoid the many wars between Burma and Thailand. For two hundred years they'd been left alone—neglected, really—and developed their economy based on fishing and farming, the products of which they consumed locally but also exported around the nation via a small port near the city. Occasionally, the military government of Burma would conscript Dawei men into building railways in order to ensure they could deploy troops to border areas in the event of trouble. Even as late as the 1990s, the slavelike working conditions forced upon them by the Burmese army were enough to push Dawei workers across the border to Thailand as refugees.

Today, Burma and Thailand pretend to be the best of friends as Burma opens up its economy to foreign investors. Industry leaders in Thailand, backed by their government, have their eyes set on turning Dawei into one of Asia's largest industrial parks, complete with a massive deep-sea port. Because the Dawei people were never asked if they wanted the project in their city in first place, many Dawei residents resent the government's trampling of their rights. A number of Dawei women's groups also object to the risk the project poses for their children. A list of their combined grievances includes the following:

1. *Land confiscations from farmers:* During the past few years, these amount to nearly seventy thousand acres of confiscated land and the eviction of tens of thousands of local Dawei from their homes. The Dawei have been promised new land and/or compensation, but so far this restitution has been distributed in a very patchy manner and at a scale way below the economic losses the people have suffered.

2. *The ruination of the Dawei agricultural economy:* The farmers in and around Dawei City grow betel nuts, cashews, durian, pineapples, mangoes, and mangosteens, both for the local economy and a little for export. Most of this agriculture will have to be abandoned as

Dawei City 2121

¤ ¤ ¤

the land becomes industrialized. Wealthier farmers bemoan the effect that industrialization will have on their income. Poorer farmers bemoan the fact that they rely on growing their own food and if they can no longer do so they will go hungry. They also do not want to suffer the indignity of being too uneducated to be employed in the factories built on their own confiscated land and unable to afford the food sold in the city's proposed new shopping centers. Facing such food insecurity, many Dawei people report that they expect to have to pull their children out of school because the money spent on school fees will be needed to buy food.

3. *The suffering of the sea:* With chemical runoff from factories and the dredging of the sea floor to allow the free movement of ships, the coastline will be altered and degraded profoundly. Not only will the project undermine fishing—replacing the current fishing port with an industrial port—it will likely decimate the coastal ecology of southern Burma.

4. *Industrial colonization:* Environmental activists see this project as a way for Thai developers to circumvent ongoing site-selection problems in their own nation as they plunk the industrial park just across the border in Burma.

If Dawei farmers are turfed off their land, Dawei laborers cannot find employment, and Dawei children cannot get a basic education, then there is the risk of a large-scale economic migration from Dawei to Thailand—much of it illegal—as Dawei residents seek better opportunities to make a living. This will potentially break up families and decimate the community.

At the moment, there is still hope. The project is only in its first phase, and the military government that so determinedly pushed the Dawei off their land has—for the time being—relinquished control of most of the nation to elected civilians, in whom there rests more optimism about getting a fair hearing.

Presently in Dawei City, there is a classy showroom advertising the project as "Dawei: The Next Singapore, Linking Southeast Asia with the Rest of the World." However, if one peers out the showroom window at the project, there are as yet only a few waterbirds perched gently on palm stumps among swamp-bound reeds.

Despite pressing ahead with evictions and land zoning in the 2010s, the whole project is liable to come to a sputtering halt for lack of investors. The developers think they have found salvation in the Japanese government, which has promised billions of yen for the project, but the human rights issues may be set to become increasingly visible as Burma's unsteady democratic state becomes more stable. Meanwhile, the financial rewards are not altogether certain, so that any future Japanese government is just as likely to pull out as it is to stay involved with the project. Already the project has been put on hold, then restarted, then slowed down, then downsized, then upsized, then downsized again numerous times.

Probably, one day in the future, the commercial unviability of the project will be plain for all to see, and Japanese and Thai investors will decide to pack up and leave in order to cut their losses. The journey to this point may take many decades, and it will depend on how far along the project has proceeded and whether the Burmese government moderates its bias against ethnic minorities, but perhaps the Dawei people can recover to reestablish their city on the basis of fishing and farming. The air will remain clean and clear, and the sea and waterways full of fish. If this happens, then Dawei 2121 will be its own preserved ecotopia.

Alternatively, however, the army might force its way into national leadership once again. Then financial salvation for the entire project would come from China, as the Chinese government seeks to expand its physical presence and economic influence in Southeast Asia. But even with the full financial might of China involved, there is likely to be delay after delay as civilians confront the power of the army, various governments wrangle deals tied to international politics, and developers push and pull against economic forces. These delays will likely cause insecurity and frustration for the local Dawei people, since they will not be able to plan for their future. Yet it will also give them time to learn how to either resist the project or adapt to it.

Even if the delays are finally overcome by Chinese developers (maybe in ten years' time, maybe in twenty years' time, maybe far in the future) and the project develops to its full glory, the Dawei locals may be in a better position than they are now to demand a say in the process. Instead of just submitting to the biased wisdom of the industrial sector and the slick promises of construction

Dawei City 2121

¤ ¤ ¤

corporations, they will either mount a resistance effort via legal challenges to force more delays or demand to have a part in drawing up the final plans for the project. Of course, it's quite possible for both of these alternatives to be combined so that the development is mandated to proceed in a step-by-step fashion only if the suggestions of the locals are adhered to at each step. In this case, here are two suggestions to make the industrial park more eco-friendly for the coming decades or for 2121:

1. *Eco-power:* All personnel involved in the project, from low-paid workers to general managers and investors, have to ride on electrogenerating pedal bicycles when they move around on-site. This provides two environmental benefits: first, the transport of personnel is achieved without producing waste gases, and, second, the electricity for the industrial park is—at least in part—produced in an eco-friendly manner, as the electricity feeds back into the electric grid.

2. *Biopower:* Everyone involved in the project must grow organic vegetables within the industrial zone and use them to provide lunch for the workers. This has multiple benefits. First, it offers proof to all that there is no contamination on the site. Second, it decreases the ecological footprint of the park's labor force. And, third, it encourages the factories to drastically lower their chemical wastes.

These suggestions, if codified into law, will not only make for an eco-friendlier industrial zone but will also drive small-scale local innovation as people work out how to plan the transportation system, the energy system, and the food system for a new Green type of Asian industrial city.

Denver 2121 ¤ Another Green Games

Denver in the early nineteenth century was just a rocky bluff in the American West that divided the South Platte River from Cherry Creek. At that time the land was part of the buffalo grounds of the nomadic Arapaho. It was not until 1858 that Denver was founded as a permanent frontier town. Its name was chosen by the land speculator William Larimer to curry favor with the governor of the Kansas Territory, James Denver. Denver had already resigned as governor by then, but the lines of communication were so patchy on the frontier that Larimer didn't know. The name stuck, and the town grew—mainly to service the miners prospecting in the Rocky Mountains nearby. Because of its proximity to the mountains, and because of its high elevation (officially it is 5,280 feet above sea level), Denver earned the nickname the Mile-High City.

Denver in the early twentieth century was a pioneer in the auto industry due to the work of the Colburn Automobile Company. By this time, Denver was the capital of the state of Colorado, and the Colburn cars produced there could be seen cruising around Colorado cities with a large brass C emblazoned on their hoods. Although the company stopped making cars by the 1910s, Denver developed over the ensuing decades to become a typical American automobile city, with hundreds of miles of paved roads and thousands of traffic lights.

Denver in the early twenty-first century is a city of more than two million people. It's regarded as a very good city to walk around and to bicycle across because the terrain is flat and many parts have dedicated pathways for pedestrians and cyclists. However, the automobile still dominates as the main form of transportation, and it's often very difficult to get from one part of the sprawling city to another without using a car.

Denver in the early twenty-second century is another city entirely. It has fought against car dependence by converting the entire metropolis into a small series of self-contained eco-towers, each providing habitation for tens of thousands of people. These megastructures are the perfect antidote to urban sprawl. An eco-tower requires about 2 percent as much land as a typical city of similar population. With little or no need for cars—or massive roadways—such urban regimes may also give rise to a less polluted landscape. Residents also hope that the decimated populations of the American buffalo can be restored as roadways and parking lots are decommissioned and turned into rejuvenated wildlands between the towers. The height of the eco-towers also allows Denver 2121 to outclass itself by becoming the Two-Mile High City.

For some, this compact, self-contained city is both eco-friendly and convenient. Others, though, would say that such towers are enormous monuments to planned societies that would envelop the individual in an abominable techno-prison. The obvious riposte is that the automobile has already imprisoned us in a technological nightmare. The tower here symbolizes all this with its gigantic upturned gas-funnel form. Although rather megalithic, each eco-tower is adapted to its Colorado setting. Encircling each tower's core are layers of horticultural plots that grow native edible crops such as corn, prairie turnips, wood strawberries, wild lettuce, and Colorado prickly pears. The types of crops cultured in these layers gradually vary with altitude, depending on their natural physical limits.

◻ ◻ ◻

Another reason Denver 2121 can be classed as an ecotopia is that it strives to reinforce its environ-mental credentials by publicly restating its twentieth-century Olympic Games rejection policy. In 1972, Denver became the world's first city to be awarded—and then reject—the honor of being an Olympic host city. The reasons for the rejection were both financial and environmental, and these are the same reasons that encourage Denver's subsequent Olympic rejection in the twenty-second century.

In the future, some of the world's largest cities will still clamber over each other competing for the right to host the Games. Decade after decade, the show gets bigger and bigger, and the environmental and social impact gets more and more drastic. By the time Dubai gets to hold the Winter Olympics in 2121 (see next city), two rapidly built power plants will be required to operate the artificial ice and snow venues, and the Dubai city government will be forced to remove a quarter of a million residents from the newly designated Central City Olympic area.

Supposedly, the Games offer host cities the opportunity to stand out in the world, but the Denver authorities know something that other city authorities do not. Many people despise the Olympics for their showy hype, their domination

Denver 2121

◻ ◻ ◻

of the airwaves, and their rampant dope-cheats as well as for showcasing a form of entertainment they find nationalistic and boring. To cater to those disenchanted people, Denver 2121 offers Escape-the-Games vacations, where global tourists can have home-stays in an eco-tower isolated from all things Olympic.

The visitors can also take part in ski eco-holidays in nearby sustainable resorts in the Rocky Mountains. These resorts forgo motorized chairlifts—instead, skiers must trek up the mountain slopes with their skis on their backs. They then slalom down the slopes among replanted native Colorado trees. Most ski resorts have a huge impact on the ecology of a local area, but here the ecology is preserved, as the skiers encounter a near-natural mountain slope rather than an engineered one.

Ecotopia 2121

¤ ¤ ¤

Dubai 2121 ☒ This Sinking Island Earth

Dubai's ruler, Sheikh Mohammed, owns Greenland. Not the big, icy Greenland northeast of Canada, but a smaller, artificial Greenland-shaped island he had built off the coast of Dubai. This artificial Greenland is itself part of an artificial archipelago called The World. From an airplane, the archipelago looks like the nations of the globe laid out on the surface of the Persian Gulf. However, from the beaches of Dubai they are just a hazy blur on the horizon. The World was dreamed up to be for the very rich only. It was planned to be adorned with six-star hotels and magnificent villas accessible only by those who own a hovercraft or a helicopter.

According to many reports, however, most of the islands of The World are eroding into the sea from which they were dredged, and the channels between them are silting up and becoming unnavigable. The high profile of The World plus its cost (some $14 billion) marks this project as a magnificent failure for both Dubai and the sheikh. By the middle of the twenty-first century, the last island of Sheikh Mohammed's dream will likely have been submerged beneath the waves, but The World's sunken, blobby traces will still be visible to any passenger looking out the window of a plane while landing at Dubai International Airport.

One century from now, in celebration of the coming Winter Olympics, the new sheikh of Dubai will set out to combat the drowned legacy that is The World's failure. On the same site, he will build five monolithic islands that, in number, recall the Olympic rings. They are designed to outlast erosion, survive sea-level rise, and host the Olympic athletes in a monumental fashion. This is Dubai 2121: Global warming doesn't need adapting to. It needs to be beaten.

El Dorado 2121 ¤ Legacy of a Lost Utopia

Every dozen years or so, a lost city is unearthed. Previously confined to legend, or to a few texts, or maybe completely forgotten, such lost cities reemerge now and again to confront us with stories of how people lived in times past. Despite being shrouded in age-old mystery, they are usually full of symbolic meaning for those living in the time in which they are rediscovered. They tell stories of past Golden Ages and delight us with hitherto unknown forms of social life. For some, they reinforce fables about apocalypse, warning us that great civilizations can end up in oblivion. And in the era of the eco-catastrophe, when the possibility of human extinction is entertained by serious minds, lost cities offer salutary warnings about demise and collapse.

In this scenario, the lost city of El Dorado is stumbled upon again deep within the Colombian rain forest, reemerging from myth into reality. In myth, El Dorado was a City of Gold, containing dazzling golden temples filled with statues made of gold and gardens full of golden flowers. This depiction isn't really too far-fetched, since the pre-Columbian cities of the Aztecs and the Incas were viewed in the same way by the eyes and minds of Spanish conquistadores in the sixteenth century.

Yet in 2121, when El Dorado is rediscovered hidden between the Andes and the Amazon, it reveals what was once a complex and unique society, but there's no gold at all within the city. Outside of the boundaries of the city, gold artwork aplenty is found submerged in nearby swamp waters, lying at the bottom of lakes, and abandoned within caves. As archeological explorations proceed, El Dorado is revealed to be a city of the Musica, an ethnic group who inhabited parts of Colombia before Europeans settled there. The Musica did not regard gold as intrinsically valuable or as a way to display their wealth, but as a means to make offerings to nature. Nature was sacred. Gold came from nature—a gift or a spirit from the sun god—and back to nature it should be given.

Florence 2121 ¤ The Germ Theory of Ideas

About seventy years into the future, when Leonardo Scimiescco is a final-year student at the Florence Art Academy, he is too wild and energetic to sit still in class for three hours a day drawing bowls of fruit. However, he does enjoy entering the "technopreneur" competitions that come to him via the design department. One of these is a challenge to design a new, eco-friendly toilet. So at a café-bar across the street from the academy, he scribbles down notes onto art paper, making a few calculations next to them while sketching out his ideas.

Advances in twenty-first-century physics make it possible for mini black holes to be created in a lab. Standard black holes, as most everyone knows, are collapsed superstars in darkest space that suck in other stars and planets and anything that comes too close to them. Perhaps mini black holes could do the same at a "mini" level on Earth. Inspired by this idea, Leonardo designs and draws schematics of a black hole toilet system whereby users can relax on a commode to do their business and be assured that their waste will disappear forever when the chain is pulled. The city would never have to manage sewage again. It would be gone, like magic, in a puff of mini-black-hole smoke.

When he submits his design to a panel of judges, some of them are appalled. They know that black hole technology cannot make stuff disappear forever. Whatever is thrown into a black hole is just transported along the space-time continuum to reappear in another place. Thus, Florentines could flush their black hole toilets only to have the waste reappear out of thin air in Rome, where it might whack the pope or the prime minister in the face. Both Leonardo and many Florentines might think Rome deserves no better, but they have to admit that black hole technology suffers from the same problem as many other technologies: the *externalization of disbenefits*, whereby one person's solution becomes another person's problem.

However, Leonardo also submits another toilet design, which combines biotechnology with information technology to create a genetically engineered "smart biotoilet" that recycles bodily waste into usable organic matter and nitrates and then packs them into small, dry pellets for convenient use in residential gardens. This toilet is also capable of a decent conversation while in use, as well as being soft and warm to the touch. Instead of cold, hard ceramic, the biotoilet feels more like you are sitting in the lap of a large, cuddly teddy bear.

Most of the panel judges don't like this design very much either, except for the head judge, who is the CEO of an information technology firm and is always seeking to invest in new IT concepts in case one of them might take off.

Leonardo ends up winning the competition and is awarded a few million dollars to develop the second design, which he totally fails to do, preferring to use the money to make amusing urban art installations throughout the city. However, others do take over the concept to make a marketable product. There is a learning curve associated with its use, however, especially in many art academies around the world. When students come back drunk and need somewhere to be sick, most of Leonardo's smart biotoilets are so disgusted with what is being forced into them that they eject it back at the offending student. Until this problem is fixed, users must be careful where they puke.

Leonardo's penchant for combining art and science, along with his eccentric innovations, are enough to earn him the nickname Leonardo II, after the Renaissance artist/scientist Leonardo da Vinci. Like his namesake, Leonardo II has an obsession with flight, and he sets out to lift Florence's buildings into the sky to float above one another, thereby increasing the available architectural space without destroying the Florentine cityscape with skyscrapers or urban sprawl. He proposes using new sturdy, lightweight plant fibers combined with helium gas vesicles that enable buildings to float in the sky in the same way that gas bubbles enable large, heavy seaweed to float in water.

Leonardo comes up with many more wild designs and theories, but usually they are confined to the drawing board. A theory of his that gains currency in Florence is the *germ theory of ideas*. The brain is generally considered to be the anatomical organ that thinks up new ideas and creates original thoughts. According to Leonardo II, though, this is not at all the way ideas and thoughts are created. Sure, the brain controls reflex actions and nerve impulses, he says, but you can hardly get a great inert lump of squishy pink meat to think up original ideas. The creative process, said Leonardo II, is the result of a contagious disease, which is transferred via *idea germs*, microscopic amoebae that live within the fine, spore-laden organic dust of many urban environments.

And so in Florence 2121, the usual cleaning and disinfecting of everything in the home and the workplace become things of the past. The machines and chemicals once used to sterilize the urban world are retired, and a healthy, dusty microbial ecosystem becomes the fashion.

One rather unsavory side effect of the growing popularity of Leonardo's germ theory is that an increasing number of students in Florence 2121 seem to spend most of their time licking the germs from the walls in dusty nooks in a desperate attempt to ingest the accumulated microbes. Florence as a whole becomes shrouded with a soft microbial dust cloud—they nickname it "the Mystique"—as people work to create an appropriate ecosystem for creativity.

Ecotopia 2121

□ □ □

Gaia 2121 ¤ The New Age City

Vila Nova de Gaia, or just Gaia, as the locals call it, is a Portuguese city with a population of two hundred thousand. It's not really a standout city by any means, its most notable feature being the many cellars where barrels of port wine are allowed to age. By the early twenty-second century, though, it has become a core Green city of Europe, serving as a well-known haven for eco-spirituality. Gaia 2121 has cathedrals devoted to nature worship and colleges run by religious eco-warrior gurus. There are also statues honoring pagan gods and a dozen communes devoted to sustainable viniculture. Although the environmentalists in Gaia call themselves Gaians, it's unclear whether this refers to the city where they all live or to the worship of Gaia, the Greek name for Mother Earth.

Over time, various extreme factions of Gaianism tend to develop. One eco-priestess named Super Nova roams around in a hazy, noisy, dreamlike state every night, screaming her abiding loyalty to Mother Earth. Another Gaian, an engineer who goes by the name of Bucky Fizz, draws up plans to build geodesic domes on the planet Mars, believing that interplanetary conquest would be the natural way for Mother Earth to give birth.

Mostly the interactions among the various Gaians are cordial, but a confrontation does erupt one evening when Super Nova drinks too much port wine and crashes into one of Bucky's domes, smashing all his gadgets while yelling that Mother Earth is greater than human engineering.

Despite their differences, most Gaians seek to recognize the intrinsic rights of nonhumans, believing that animals and plants that share the world with us have as much right to exist and be cared for as any member of the human species. The practical upshot for the city is that by 2121 its factories and ports are bought out, appropiated, occupied, or burned down, unless they are patently determined to be respectful to nature. The rest of the country doesn't really care that much because Gaia has always been a pretty boring place, and at least all this Green stuff adds a bit of color to it.

Goa 2121 ⌗ Eco-nudism on the Sea

Goa is a former Portuguese colony on the Indian coast of the Arabian Sea. From 1510 to the mid-twentieth century, Goa was largely independent, though protected by the Portuguese government. In 1961, it was annexed to become a state within the Republic of India. The Portuguese legacy is noticeable in the architecture as well as through the large Catholic community that persists.

In the twenty-second century, the Goans may be feeling the need to put some distance between their state and the rest of India once again as a result of some of India's objectionable environmental decisions. For example, if we project some of India's ongoing development impulses into the future, it is likely that the following will be forced on or will impact Goa:

- The lack of commitment shown by the Indian government to slow down the nation's greenhouse gas emissions
- The prospective construction of experimental thorium nuclear plants in Goa
- The selling of Goa's forests for timber exports
- The encouragement of laborers from all around India to come and work in Goa

- The development of an aluminum smelter near Goa's Western Ghats nature reserve

A major problem is that Goa is suffering dire global warming problems, but the Indian government does not seem to want to work at mitigating them. Indeed, because Goa is the wealthiest state in the country, the Indian government prefers to levy an extra tax on Goans in order to fund climate change adaptation measures. The Indian leaders don't seem to want to work to halt climate change and instead prefer that everybody just put up with it and adapt as best they can. Added to this is the fact that many Goans do not want non-Goans to come and settle there, since they feel that an increased population will only make problems worse. So Goans will probably campaign, ever more dramatically, to be independent again.

Whether or not full independence is achieved by 2121, Goa will be afflicted by global sea-level rise. In this scenario, the growing coastal town of Goa Velha has had to adapt physically to the rise in the sea level and the erosion of land. Families of Goa Velha are now occupying prefab houses perched on sturdy stilts. By 2121, every new school and health clinic in Goa is built in this same way, and the predominant form of transport between them has become the traditional dhow.

As the sea level has risen and Goans have watched parts of their state disappearing, the city government—as well as the Catholic Church—has set about imploring people to adopt an adaptive, eco-friendly lifestyle so that the climate-related problems are not exacerbated. This is likely to have unexpected consequences, since some individuals will decide to embrace a simpler life, giving up cars, luxuries, and also, maybe, their clothes.

According to nudists and naturists of today, the benefits of not wearing clothes include:

- Encouraging a lighter ecological footprint by eschewing the resources needed to make clothes
- Ejecting clothes-based stratification in society
- Encouraging sports and fitness and hygienic bathing
- Encouraging a healthy diet (for the sake of a more attractive appearance) and a healthy social life
- Encouraging tourism and international solidarity within a global community

The new architecture of Goa Velha 2121 in itself does something to promote an initial burst of family nudism, since the homes are far enough apart that the residents can forgo clothes without annoying neighbors or passersby. More and more, though, social nudism becomes a trend, and Goans begin to visit beachfronts, then entertainment centers and shopping venues, without clothes.

In the years between now and 2121 AD, there are likely to be many supporters and many critics of the practice of social nudism, including some emerging from sections of the Indian government and the Goan Catholic Church. However, as Goa becomes famous for social nudism around the world, its economic value to Goa becomes apparent, since naturist tourists flock to the state, bringing money and an ever-more-liberal acceptance of the practice. The Goa state government is soon vexed about whether to discourage, ignore, or embrace the practice. Many arguments ensue, and successive governments will undoubtedly have different attitudes about it. Some will suggest that nudism should remain a private practice confined to personal residences, but this does not stop tens of thousands of naturists from happily wandering around Goa conducting their daily affairs without wearing clothes.

Apart from the beach life and hippie culture of Goa (which is readily apparent even today, in the early twenty-first century), one of the reasons social nudism finds acceptance here is its historical precedent. In the fourth century BC, Alexander the Great traveled to India and found wandering groups of naked holy men, whom he dubbed "the naked philosophers." Christianity, as well, has been more accommodating of nudity in times past, so even the Catholic Church might be able to tolerate it, especially when it realizes that many naturists are not only strongly spiritual but also committed Christians. Nudism and naturism also advance the values of vegetarianism, teetotalism, and yoga, which are already well respected in most parts of India.

By 2121 AD, social nudism in Goa has become fairly well united with the environmental aspirations of the Goan state. Besides that, it is also becoming a growing part of a cultural identity that makes Goa different from the rest of India, reinforcing the political aspirations of those Goans who campaign for independence.

Gongshan 2121 ¤ A City Not Dammed

The Salween River valley in China is a landscape of rare beauty, often called the Grand Canyon of China. It is the last large, free-flowing river in the nation; all the others have been dammed, diverted, or disappeared into dryness. The Salween begins on the Tibetan plateau and winds through China, then Burma, then along the Thai border, and into Burma again before ending in an estuary on the Andaman Sea. Along its banks are five million people, many of whom are subsistence farmers who depend on the river's water and its fish.

Today, as China proposes a series of dams on the river, the people near the Salween know that it could soon be forever changed. They also fear a huge disaster if a dam breaks during an earthquake or a flood. For these reasons, and due to difficult negotiations with Burma and Thailand downriver, the dams have been put on hold—for now.

Let's fast-forward to the next century. On the first day of January 2121, the residents of Gongshan city are told to evacuate their homes and leave the city immediately. The dam project has finally begun, and their small city is to be submerged. At first, they refuse to budge. They were scheduled to meet with lawyers on January 2 to map out a way to stop the dam. It seems the evictions were begun early in order to preempt the meeting. Meanwhile, construction of the dam has begun upriver.

For one year, the residents refuse to leave their homes, afraid the buildings will be demolished if they step outside. Supporters and relatives from other river towns come to deliver food and offer moral support. The government knows forced evictions might look bad, but officials are eager to push ahead with the project. However, on December 31, 2121, one day before the order comes for the Chinese army to move in on the Gongshan protestors, a week of heavy rain causes a massive landslide right where the dam is being constructed. The landslide crushes the machinery and infrastructure under a mountain of rocks but leaves the Salween running free. Nobody is hurt, but the project is in ruins and has to be abandoned. Gongshan survives in peace, and the residents emerge from their homes to thank nature for its help.

Graz 2121 ¤ Go Green or Go Away!

Graz is the capital of the Austrian state of Styria. Graz today is a Clean-Tech hub in Europe, a sector that includes renewable energy, recycling technology, and zero waste production. The sector earns hundreds of millions of dollars per year for the city, and it is growing by leaps and bounds. All this means that Clean-Tech is becoming highly influential in Graz, firmly establishing itself within the city's ethos and identity. In coming years, anybody who works against Clean-Tech—by not using it, for example—may be called out as being bad for the city.

Back in 1599, Johannes Kepler, one of the greatest mathematicians of all time, who worked out how the planets move around in space, was living in Graz and working there as a teacher. At that time, a cultural war was being waged all over Europe between Catholics and Protestants. The leaders of Graz were Catholic. Kepler was a Protestant. In 1600, during a crackdown against Protestants, Kepler was forced to leave the city.

Graz today, and Graz in the future, is a pretty city with beautiful suburbs set in forested mountains. The residents want to keep it that way. The liberals among them are inclined to do this through environmental policies. The conservatives, though, are inclined to do it via policies that focus on cultural identity and strict immigration control. In the twenty-second century, we find these two value systems fighting for the future of Graz. The liberals want every organization in the city—be it public or private—to be eco-friendly and use Clean-Tech. The nationalists, on the other hand, want every organization to be more "Austrian." In 2121, they strike a deal and forge a new policy—one as strict as any in Graz's cultural wars. Everybody in Graz must live a Clean-Tech life or else be banished from the city. Thus, if you are found to pollute the cityscape in any way or if you are merely found to be littering, your permit to reside in the city is permanently revoked. The nationalists of Graz believe that foreigners cannot possibly be as clean as Graz's new law will require them to be, and they expect that these new immigrants will soon violate one rule or another and attract the uncompromising attention of the border police.

Greenville 2121 ✠ Home of the Sunflower

Greenville, South Carolina, is a city with a delicate little secret. Its wooded peri-urban roadways are home to the rarest and most endangered sunflower on Earth, Schweinitz's sunflower. Only twenty small pockets of the plant exist, each containing just a few dozen specimens. Not that many people really care about this situation except a few botanical enthusiasts—that is, until, later this century, when a local firm launches its new kit-form off-grid home, dubbed the Sunflower Home. Because they are easy to construct, inexpensive, and self-sufficient, Sunflower Homes sell like hotcakes. The roof is paneled with a solar array to provide for all energy needs, and water is collected from the roof and channeled to holding tanks in the walls.

People always wonder why they're called Sunflower Homes, since they look nothing like sunflowers. To explain, the designers talk about heliotropism, the sun-tracking behavior of various types of plants. This behavior allows plants to maximize their collection of solar rays as they turn the faces of their leaves and flowers to slowly follow the sun in its daily journey across the sky. The most famous of the heliotropic plants are the sunflowers. If you trace the path of a sunflower bloom as it moves across an arc to follow the sun, then fashion that shape into a residential building, you end up with the shape of the Sunflower Home's roof.

The popularity of the Sunflower Home by 2121 works in three ways to conserve the remaining pockets of Schweinitz's sunflowers. First, it increases public knowledge of the plight of the rare sunflowers. Second, the designers invest a good proportion of their profits back into conservation programs aimed at protecting rare sunflowers. And, third, because the sunflower home fits gently into the landscape, it is much lighter on the environment than conventional buildings.

Hanoi 2121 ¤ Utopia through Neglect

Over the course of four thousand years, Hanoi was built, then lost, and then built again as a series of citadels among a network of waterways and lakes. These citadels were taken and retaken by a long line of successive kingdoms and dynasties, both Vietnamese and Chinese. The first citadel site still exists in a spot called Co Loa, rising slightly above the northeast end of the sprawling city of six million.

The waterways and lakes of Hanoi today are detestable: garbage-ridden, unsanitary, and sometimes toxic. They have become major health hazards as reservoirs of chemical pollution and biological disease. As the Vietnam government seeks large-scale foreign investment to create industry throughout the city, the waterway pollution is only getting worse. While the government celebrates the growth of its capital, many Hanoi residents continue to have little access to either clean water or a sewage system.

For this reason, in the future many poorer urban communities, including the vast numbers who have migrated to Hanoi from the countryside, become completely disenchanted with the communist government, declaring openly that they never get anything from the new industrial developments—no jobs, no schools, no hospitals, and no health care. Surely, if there's anything that a communist government should provide it is universal health care and free education, but Hanoi's poor do not receive even that. They just have to take care of themselves as best they can.

Then, in 2019 and 2020, huge floods sweep through the city, drowning many people—mostly children, the sick, and the elderly—and devastating entire neighborhoods. The response of the government is not to do much to help the flood victims but, instead, to construct a grand new federal center on Co Loa, so that important officials don't get wet the next time the city floods.

You might think that by this time the poor people would have had enough—that they would march upon Co Loa and demand that hospitals, schools, and power stations be built on the citadel, not government offices. But, no, they prefer to be neglected. For they know neglect brings relief from paying taxes and compulsory military service, and their communities can invest the resources they save into making a flood-tolerant cityscape that mimics the water-based agriculture of their hinterland. Here, in Hanoi 2121, the government and the poor people live in blissful disunion with each other.

Havana 2121 ⌗ The Art of Car Maintenance

Havana, the largest city in Cuba, has been forced into becoming quite eco-friendly by dint of international relations. In 1959, Cuba's foremost trading partner, the United States, launched an embargo against the island nation when Fidel Castro took power for the Communist Party. Within a very short time, the Soviet Union stepped in to become Cuba's new foremost trading partner. However, when the Soviet Union collapsed in 1991, so did a huge amount of Cuba's trade.

The massive barriers to trade erected by the US meant that Cuba couldn't import automobiles or the oil to run them. The main solution, still in evidence today, was to skillfully maintain the old cars that were already in the country and to conserve oil. The sudden demise of Soviet trade also meant Cuba couldn't buy fertilizers and pesticides or feedstock for animals. The main solution to this problem was to reset agricultural industry to be organic and more plant-based and to encourage city dwellers to make urban gardens. The latter has resulted in more than thirty-five thousand hectares of land still being used for urban agriculture in Havana today.

Now Cuba is on the cusp of reintroduction into the global trading market, and the United States is set to become its major trading partner once again. In the coming decades, there's bound to be a battle for the future of Cuba: should it be industrialized and trade in manufactured goods, or should it become even Greener somehow? In a way, this debate will benefit the Communist Party, since the focus will not be on *democracy versus dictatorship* but instead on *industrial socialism versus Green socialism*.

Havana 2121 has settled on a program of Green socialism with isolated pockets of foreign-funded industry. In this way, if there is any sort of crisis, the government can simply blame a specific foreign company and banish it from the country.

Houston 2121 ¤ The City as an Ecosystem

Ecotopia 2121

◻ ◻ ◻

Houston 2121 is the world capital of *industrial ecology*. Here, all the industrial elements of the city are integrated into an intricate whole so that the waste that comes from one factory is used as a resource for another factory. In fact, in Houston 2121 this is done so well that there are very few wastes and each by-product, be it solid, liquid, or gas, is recycled by other factories over and over again using minimal energy and producing zero carbon emissions. This is the perfect industrial ecosystem. Even the waste heat can be used to power the offices of the industrial bosses.

However, there's nothing natural about it—no gardens, no parklands, no greenery. And every person in Houston seems to be a cog within a vast machine landscape. Here, the ecosystem is artificial; water, oxygen, carbon, silicon, methane, and ethane are all cycled and recycled with elegant energy efficiency and little waste, but there's nothing alive down there save for a few human technicians. For industrialists, the idea of *indus-trial ecology* is attractive, since it pretends that our vast industrial civilization can be converted to an eco-friendly state if we just find technical solutions to the problem of recycling materials and energy. However, the *ecosystem* concept is a machine metaphor applied to natural settings, a vision of nature more loved by engineers than ecologists. So if you start off with the idea that you want to turn the city into an ecosystem, you'll end up getting not a living community but a machine community.

Two other problems also cast doubt on Houston 2121 as ecotopian for anybody except an industrialist. Namely, some very rare resources have to be transported from far away to keep the process going, and second, some products cannot be made without producing a waste that is so toxic it can never be used by another industrial process. Petrochemical products, a mainstay of the Houston economy, may very well belong to this category.

Karachi 2121 ⌧ From Megacity to Mini-cities

Karachi, the twenty-million-strong megacity of Pakistan, has severe problems with the silting up of its harbor and waterways. Transportation and the marine environment are both so grossly degraded by silting that each is being brought to the point of dysfunction. The silting is a natural process, for the city's rivers—the Indus, the Malir, and the Lyari—bring many tons of soil to the coastal waters every year, but the situation is made exponentially worse by the illegal dumping of garbage, deforestation, and human-created soil erosion. In this scenario, Karachi 2121 attempts to deal with these issues together. Dredging machines dig up and concentrate silt from Karachi's harbor and build it into mounds upon which new mini-cities can be built.

How is this financed? Partially through government investment, partially through private investment, and partially through compensation provided by the signatory nations of climate treaties, which are legally obliged to reimburse those poorer nations adversely affected by climate change. Mostly, though, these mini-cities are built by people themselves, predominantly the urban poor, since they are the people actively looking for homes. They seize the opportunity to start building as soon as the new islands emerge from the water.

Currently, in the twenty-first century, the slum communities of Karachi are often criticized for clogging the waterways with houses built on wooden stilts, which is said to exacerbate flooding. But the success of these communities in colonizing silted-up areas with few resources, and the way they solve their own problems, should impress those in local government.

In the twenty-second century, these communities are given permits to construct a mini-city on a test island. They are given no funds, no resources, no technical help—only the legal indemnity against prosecution for occupying newly emerged land.

By 2121, a whole series of reclaimed island cities is colonized in the same way. The sustainability of each mini-city will always be of primary concern to the residents, since the physical limits of these cities discourages the accumulation of physical assets. Instead, everybody on the new islands agrees that any profit will be reinvested into shared civic projects such as eco-fishing, wave energy, and environmental education, thus allowing them to develop richer communities in conjunction with their nonmaterialistic lifestyles.

Katun 2121 ¤ City of the Glacier

The Altai are the people surrounding the Altai Mountains, a high range that marks the border between Russia, Mongolia, and China. They've inhabited this area for over one thousand years, since their ancestors moved from other areas in Western Asia. Today, in the early twenty-first century, there are three cities with the name Altai located within three provinces of the same name, one in Russia, one in Mongolia, and one in China. In each separate Altai province, the Altai people have been pressured, regulated, assimilated, and outnumbered by the dominant Russians, Mongolians, and Han Chinese. Altai ethnic nationalism in turn has been suppressed in a number of ways: by denying they have a common heritage, by dividing them from each other, and by disallowing their Burkhanist religion.

The Altai have also suffered from environmental injustices, their homelands having been pitted with mines and sullied by dirty factories. Maybe, in the coming century, they will keep dealing with this as best they can. Or maybe, as they detect domestic discord that weakens Russian, Mongolian, or Chinese management of the area, they will band together to mark out an independent territory.

In this scenario, Altai people organize to push into the mountains near the border between the three countries. Here they begin to build a new city, calling it Katun, named for one of their sacred glaciers in the Altai Mountains. (Katun City isn't actually anywhere near Katun glacier, but that works to confuse policing agents about the city's whereabouts.) The city is so deep in impenetrable mountains, and so shrouded in snow and an ever-present mist, that it's invisible from the air and inaccessible by vehicle. If any one of the three nations sends in an anti-Katun brigade, the Altai know they can walk a few hundred yards to escape across the border to the next nation. They also know it's extremely unlikely that Russia, Mongolia, and China could cooperate to advance on Katun at the same time.

The Altai Mountains may seem like far too challenging a place to found a new city, but it is one of the few areas of the world acknowledged as being unaffected by climate change. While the Russian and Mongolian grasslands shrivel for lack of water, and China's supermonsoons occasionally flood the land with too much water, the Altai people will live peacefully in a predictable natural environment.

Košice 2121 ¤ The Blue River City

Košice, the second-largest city in Slovakia, once had the Hornád River flowing through it near the city's center. The river descends from the Carpathian Mountains through the city and out again across the Hungarian border to the south. Once upon a time, Slovaks happily caught fish from the river and served them at the dinner table.

Back in the 1970s, when Slovakia was part of Czechoslovakia, the Communist government decided to have the Košice stretch of the river filled in and paved over and a ring road built in its place. The water flow was diverted away from the city's center past factories and into the suburbs. The diversion had a great impact on the water creatures downstream, which are probably only just now recovering. The gaping concrete channel where the river once flowed is now painted with wavy blue lines, themselves rather faded and peeling, serving as a pathetic reminder of where the river had once flowed.

Košice is a steel city; the economy is dominated by a huge steelworks on the outskirts. In the past, the steelworks contributed to the pollution of streams that flowed into the Hornád. Parts of the river are still very pretty and tree-lined, but it has many dilapidated sections, and most parts are devoid of the once-diverse animal life.

The commercial viability of the steelworks is actually quite dubious, but the Slovak government is very keen to keep it afloat in the future with subsidies, tax incentives, and lax environmental laws. One result is that some Košice suburbs are now bathed in sulfurous smog and contaminated with heavy metals almost every day.

Eventually, in the early twenty-second century, the steelworks is set free onto the market without subsidies. The very next year it closes down, and the environment of Košice rapidly improves. Enough steel remains in storage to keep people employed in constructing the new blue city envisioned here. And by 2121, the river is allowed to flow through the city center once again, bringing coolness and freshness to Košice, as well as healthy waters good enough to resume fishing.

Ecotopia 2121

⌷ ⌷ ⌷

Lanzhou 2121 ¤ The City That Moved Mountains

Long ago, Lanzhou, set among a hundred mountains near the Yellow River, was the home of the Qiang. In many places, the Qiang planted white stones around their granite homes to honor their mountain god while farming buckwheat as their staple food near streams and rivers. More than a millennium ago, Lanzhou was invaded and overtaken by the Chinese Han.

Lanzhou in the early twenty-first century is a city of four million people, which is in the midst of expansion, transforming itself into a megacity on the scale of Beijing or Shanghai. The plan is expressed in a clumsy government motto: *Rebuild Lanzhou: Strong Industrial City, Ecological Green City, Many Lakes City, Modern New City.* The residents don't have any say in the matter, so it's been a nice, easy sell—especially since Lanzhou is currently overindustrialized and very polluted, so a nod to any kind of Green doesn't go amiss.

However, the manner in which Lanzhou is megasizing is grossly antiecological. Many of the mountains around the city are being blown up or bulldozed and their rocky remains used to fill in the valleys. Some scientists have expressed their concerns about this enterprise, pointing to risks of landslides, sinkholes, floods, and massive pollution of both the water and the land. But the mountain clearing goes on.

By the mid-twenty-first century, it is probable that there will be massive pollution and huge land collapses, and the Yellow River's very course may also change. Parts of Lanzhou may suffer near total destruction. Many Chinese will be forced to leave because they will have no homes and no jobs, only a pockmarked city.

By 2121, though, things have likely settled down. Some mountains have survived, though they are carved into bizarre shapes, and they may be stable enough to live on. With the Chinese leaving for other places, the Qiang return from other parts of China, and they set about rebuilding their homes and planting buckwheat. They also replace the sacred white stones to honor the mountains they live on.

La Paz 2121 ¤ City of Peace, City of Bogs

La Paz is set on a high plane in the Bolivian Andes, sixteen thousand feet above sea level. It is so high and the air so rarefied that water in La Paz boils at 192 degrees Fahrenheit.

La Paz today is suffering from chronic drought. In the past, most of the city's water has come from six big glaciers in the Cordillera Oriental mountain ranges, but these glaciers are receding due to climate change, and their yearly quantity of meltwater is dwindling. One glacier, the Chacaltaya, disappeared in 2008, and the snows and rains feeding the others are becoming less frequent.

Already as much as one-quarter of La Paz's population does not have ready access to water. Faced with this shortage, and with a population boom as people migrate from the countryside to look for work, the governor of La Paz is contemplating moving millions of La Pazians across the country to the wetter Amazonian parts of northeastern Bolivia.

Another solution is the rehabilitation and preservation of waterlogged peat bogs that once surrounded the city. The peat bogs acted to drastically slow down the flow of water from the glaciers to the city, meaning water was available all year round. For fifty years or more, these bogs have been dug up for mining or eaten away by sheep. The mountain meltwater now zips speedily through the landscape and through the city, leaving La Paz with little water for many months each year. Sometimes, when rain showers hit near the city, the water rushes too fast into La Paz and causes disastrous flash floods and mudslides.

In this scenario for La Paz 2121, both sheep farming and mining are completely phased out and farmers go back to the tradition of herding llama, an animal native to the area, which is much gentler on the bogs than sheep. With the livelihood of countrysiders restored, there is also less migration into the city, further helping to mitigate the water shortage.

Lazika 2121 ¤ City of the Frog Laws

The Georgian government plans to build a brand-new city of a half million people completely from scratch on a swamp beside the Black Sea. It's been named Lazika, the ancient Greek name for this part of Eastern Europe.

Building cities on swamplands certainly has historical precedents: Rome, Bangkok, St. Petersburg, Rio de Janeiro, New Orleans, and Dublin were built on swamps. Sometimes this was because the swamp served to deter invaders; other times it was because swamp zones happened to be in useful locations for maritime trade.

In the twenty-first century, the ecological value of swamps is slowly being recognized by urban planners worldwide. Swamplands in cities improve water quality, reduce flood damage and erosion, provide recreation for city dwellers, and serve as habitat for wildlife, including economically valuable wildlife. All these benefits are provided free for the city by nature, but if we had to engineer them from the ground up they would cost billions. Currently, the Lazika plan is to wipe clean away these ecological services to build the new city—a plan that would also endanger the pristine habitat of the Kolkheti National Park nearby.

An alternative plan, Lazika 2121, seeks to develop a smaller-scale coastal city that utilizes the swamp to provide eco-services for various traditional crafts of Georgia. Any new heavy industry, though, must be built at least five miles inland from the coastal swamps. Here, too, the environment is protected via the Frog Laws, which state that only industries that do not harm the many species of happily chirping Georgia lake frogs are given permits to operate. Every six months, the noise levels of the frog-inhabited lakes and streams surrounding the inland factories of Lazika are monitored, and if they are found to be too quiet, the owner companies must restore the frog habitat by the end of the next half-year or else forfeit their profits.

Leuven 2121 ⌘ Cabbage Capital of Europe

Oliver Cromwell, Vladimir Lenin, Nelson Mandela. Why do no photographs exist of these men consuming cabbage? In the eco-friendly age of the twenty-second century, it'll probably be considered impolite to insult any Green organism, but let's face it: for children all over the world, cabbage is still likely to be regarded as one of the vilest foodstuffs in the entire universe. Just a small whiff of it bubbling fetidly away in a cooking pot will likely make future kids gag, as it does today.

Despite these facts, cabbage seems to remain a popular dish in northern Europe. One theory for this strange circumstance is the use of cabbage as a parenting tool. Most children would happily go through the whole week obeying their parents' every whim just so they could be allowed the one misdemeanor of not eating their cabbage at dinnertime. A related theory suggests that the children who positively refuse to eat their cabbage go on to become great revolutionaries. Thus Cromwell, Lenin, Mandela, and the significance of the fact that no photographs exist of these men eating cabbage.

But, of course, as people mature, a taste for cabbage is often acquired. In the Flemish city of Leuven 2121, the citizens have turned cabbage into an art form, as the city promotes itself as the Winter Vegetarian Capital of the World. Every day of winter is designated as a Veggie Day, on which restaurants and schools serve vegetarian meals. Cabbage is the queen of the vegetables in this season and a darling among "eco-foods," since cabbage production does not require heated glass greenhouses, it quite likes cooler climes, and if you don't mind eating it with a few moth bites in it, it doesn't require pesticides to grow. Cabbage can also be grown wild to add to the biodiversity of the Leuven cityscape.

London 2121 ¤ Where Do the Children Play?

At the height of the hippie movement in London, during one of the city's economic boom times, musician Cat Stevens, now known as Yusuf Islam, wrote the song "Where Do the Children Play?" in which he reflected on the idea of progress, including in his hometown of London. In the song, he laments the hubris of economic and technological development, especially in the form of skyscrapers, at the expense of the simpler things in life, such as places for children to grow and play.

Many years later, the United Nations set out to describe the features of what they believed would make a "child-friendly city." Such a city, they declared, would allow kids to: (1) meet friends and play, (2) walk safely in the streets on their own, (3) enjoy green spaces with plants and animals, and (4) enjoy equal citizenship within their city.

Future governments of England will probably like to make a big fuss about these laudable goals in front of various international forums, but there's a real risk that, in the London of the coming decades, urban children will be worse off than they are today. Sure, the children of the rich will still be able to access private spaces to play with their toys, but both rich kids and children of less-well-off families will find the cities of the future an unnavigable confluence of uncrossable roads and overdeveloped private buildings.

In the London of the late twenty-first century, it might be that public parks have been ripped up to make way for corporate skyscrapers while inner-city schools are allowed to degrade and decay. Sometime during a future economic downturn, the government in Westminster will likely declare that it can't afford to maintain public parks anymore or keep paying for all the state schools. By 2100, life in London may lurch toward the Dickensian when London kids line up for five hours on the first Monday of each school term to collect vouchers that afford them subsidized classes in rundown schools under the instruction of poorly paid and poorly trained teachers.

The quality of life for pensioners, who do not begin retirement until the age of seventy-five now, has also plummeted as cutbacks in pensions take effect, the government having declared that it can't afford to pay for all the pensioners either. These two groups, the children and the elderly—both considered the most honored and protected segments of society—become marginalized. Until, that is, a series of events brings them together in 2121.

On a summer afternoon in 2121, the reigning English monarch—let's call her Queen Maria—takes an evening walk with her granddaughter, Princess Morag, to care for the roses in the Tower of London garden. "What's inside the castle, Grandma?" the child asks. Grandma, the Queen, explains that inside there are just a load of jewels and such things. "What are they doing there, Grandma?" the girl persists. And Grandma continues to explain that they do nothing but sit about gathering dust. "Oh, why can't we use them to grow more gardens and flowers, Grandma?" the girl pleads. The Queen laughs and suggests that people do not appreciate flowers and gardens anymore. To this, the child says, "You can teach them to, Grandma, just as you've taught me."

This conversation would have drifted away with the scent of the roses had it not been for two other events happening at around the same time in London. First, as a response to UN criticism that England has not done anything to fulfill its obligations to children, the London Council grants voting rights in its elections to all children ten and older. Second, Grey Power protests, comprising nearly a million angry pensioners, close down the middle of London for months on end. Because it is summertime—and school is out—the grandparents among the protesters bring their grandchildren along, and as a united front they also stake out a right to free lifelong education.

Because the protesters have closed the streets to traffic, the city's air quality improves dramatically. This does not go unappreciated by most of the other residents of London, nor by the hospitals, which notice far fewer admissions related to respiratory problems. The Grey Power protests then assume a Green hue, as environmentalists gather with the grandparents and grandkids to support their cause.

With the government holding fast to budget cuts, the protestors must pursue a radical solution. They convert eight square miles of central London into a massive eco-village, transforming unoccupied offices into residential buildings, sowing gardens on street corners to grow their food, and setting up small, sustainable businesses to trade among themselves. They also invite all children of London under age twelve to come and learn these skills for free as day students, sending them home to their parents in the evening.

London 2121

Thus, London 2121 is born, a place where children can learn and play in a Green environment and in safety.

The plan is that no one between the ages of twelve and seventy-five may enter London 2121. Also, no authority figures are allowed to enter London 2121: no parents, teachers, police, or politicians. Traffic wardens are welcome, though, and are popular figures, for they jauntily ensure that no cars can drive in the village. To make sure the area is secure, the Queen donates physical portions of the Tower of London to surround London 2121 with an impenetrable wall.

Surprisingly, the rest of London leaves them to it, for a combination of reasons:

- The Queen and her young princess, Morag, are enthusiastic about the whole plan, since both of them are within London 2121's age limits (and they would welcome some help with the rose garden).
- Many of the children in this part of London have the right to vote in the local government elections, and they all vote that the London 2121 plan is perfectly legal and logical.
- The whole arrangement ends up providing free day care and free education for tens of thousands of families while saving the government a load of money, since the large financial burden entailed by pensions and education has been relieved.
- The children trained in London 2121 acquire real-life skills from a vast pool of superexperienced and enthusiastic teachers, such that their education ends up being highly valued.

As well as all this, there's a centuries-old precedent of a similar "in-house" pension scheme just across the River Thames in the Royal Hospital Chelsea, where scarlet-coated Chelsea pensioners, retirees from the army, are given free room and board and free health care for the rest of their lives. The hospital was set up by King Charles II in 1682, and Queen Maria will be darned if she can't at least equal that accomplishment during her own reign.

Los Angeles 2121 ⋈ The Return of the Streetcar

During the 1940s and 1950s, the many streetcars of Los Angeles were systematically bought up and then closed down and dismantled by a group of conspiring auto companies led by General Motors. In Los Angeles 2121, streetcar networks make a big comeback as a major transportation resource due to the closure and redevelopment of the highways. Instead of being paths for cars and trucks, LA's highways are converted to vegetated greenways for pedestrians and cyclists. The highways also act as a network of ecological corridors, connecting populations of wild plants and animals around Los Angeles that would otherwise be isolated.

In this scenario, cars are confined to the role of filler in new Green, high-density housing, an architectural style that counteracts the problems of urban sprawl and encourages more enjoyable commuting experiences. But how can streetcars and vegetated walkways possibly serve a city of five million? First, since global warming will likely degrade LA's ideal climate, it becomes a less attractive place to live, so there are bound to be fewer Americans choosing it as their preferred home. Climate change doesn't alter the close proximity to the beach and mountains, but it does pose four tangible threats: (1) the summers will probably grow much hotter, (2) the air will probably be much smoggier, (3) there will probably be many more wildfires, and (4) there will probably be much less water.

The extra expense of contending with all of these adversities will likely impoverish the public purse as well as the finances of private landowners. Those who do stay in Los Angeles in the early twenty-second century will have the opportunity to try out the green walkways, whereupon they'll find their commute both much cheaper and much more pleasant. Not only will people be happy to be rid of the junkscapes that an automobile city forces upon their lives along with the accompanying pollution, accidents, traffic jams, and the stink and noise, they will also be wealthier because they will not have to spend so much money to buy, maintain, and insure their own cars.

Macau 2121 ¤ Gambling with the Environment

Macau's location on the Pearl River estuary makes it enormously vulnerable to a rise in sea level. Over the course of three centuries, the city has sought to reclaim land from the sea to allow for urban growth. Yet if the sea level rises by just a few yards, as is currently predicted, these reclaimed zones are likely to be submerged again or eroded during high tides and storm surges. The land upon which the city sits is also increasingly vulnerable to subsidence because of increased groundwater extraction. Macau is sinking. The seas are rising.

How is Macau to deal with this situation? Since Macau is one of the gambling capitals of the world, it should be sensitive to the value of calculating future risks. Sometime in the future, these risk calculators are bound to advise Macau's administrators and business leaders that drastic adaptive measures will need to be implemented to preserve the profitability of the gambling and tourist sectors. Risk assessors in the insurance industry are also likely to have an impact, since their risk advice about future potential environmental change will suggest that the casinos, the hoteliers, the tourist agencies, and the airlines will have to pay more and more each year for insurance.

To deal with such profit-eroding environmental change, this scenario for 2121 depicts the casinos of Macau as condensed and contained with one circular gambling zone surrounded by a huge walled barrier to protect it from sea-level floods and storm surges from typhoons. The rest of the city then has to be rebuilt on the nearby hills of Macau.

If the climate continues to warm during this century, by 2121 the malaria-infected mosquitoes of Southeast Asia will have expanded their range northward to Macau. Therefore, the barrier will be electrified in an attempt to protect the gamblers from the mosquitoes. Really, the barrier will do nothing to halt the root cause of global climate change, but it will allow the casinos to survive long enough to adopt Green mitigation measures such as switching to new, low-power gambling machines. Also, since Macau's casinos are often thought to be the playground of corrupt Chinese officials, it would benefit China's reputation to no end to reinvest some of the profits into conserving China's landscape.

Madrid 2121 ¤ Decapitating a Capital

Like many capital cities, Madrid is an administrative hub, a business hub, an industrial hub, and a cultural hub for a whole nation. This profusion of "hubness" attracts a great, continuous flow of people and money into the city to the point that Madrid is said to have become "overdeveloped." Some say it has become overpopulated, too, with a metropolitan area of seven million people, all driving too many cars on too many highways and throwing away too much trash.

Madrid's journey to overdevelopment began in the sixteenth century, when King Phillip II moved the Spanish capital from Valladolid. During Phillip II's reign, as with his successors, Madrid also became the capital of the entire Spanish empire, with colonies in the Americas, Asia, and Africa. The Nationalist period of the twentieth century was an attempt to continue the process, but by then most of Spain's overseas territories had fought for and gained their independence.

Today, Madrid still resists relinquishing any more political power to other regions of Spain. In the future, as Madrid works to prosper well into the twenty-first century at the expense of other Spanish cities, it will continue to build great boulevards and highways, all made of concrete and asphalt, materials unable to absorb water. The storm water sweeps across roads in small floods, washing trash and pollution into local lakes, streams, and rivers.

The solution, which is manifest in Madrid 2121, is twofold:

First, the Madridistas look to curb the use of concrete, demanding that one-half of the city's surface remain uncovered.

Second, many regions in Spain want out of it: Andalusia, Asturias, Aragon, the Basque country, Canarias, Cantabria, Castile, Catalonia, Galicia, Islas Baleares, and Malaga are all campaigning for independence. The only way for Spain to remain intact is by devolving far more political power from Madrid to all the other cities in these regions.

Once political power is distributed, economic and financial power is also spread around the nation, letting each provincial capital spend its taxes on projects of its own choosing. Madrid is thus released from constant overdevelopment and is able to de-develop to a more sustainable level.

Malaga 2121 ⌘ The Slow City

As the devolution of power from Madrid to the regions occurs in twenty-second-century Spain, each regional capital takes pride in highlighting a unique identity. Malaga, a city of half a million people on the Mediterranean coast, chooses to become a flagship city of the Slow Movement. The Slow Movement is a cultural revolution against the notion that faster is always better. The Slow Movement aims to forgo speed and quantity within everyday life and, instead, swap these in favor of enjoyment and quality. Time is turned from an enemy into a friend.

In keeping with the ideals of the Slow Movement, Malaga opts for a gradual journey to becoming a Slow City—aiming to wander gently toward Slowness over a hundred-year period. Eventually, though, by 2121, three aspects emerge to make Malaga a Slow City exemplar:

1. The Slow Food movement in Malaga 2121 aims to preserve the city's cultural cuisine and, in so doing, to preserve the traditional crops and farming associated with the countryside around the city. The fish-salting industry, the original trade of Malaga some twenty-five hundred years ago, is also resurrected.

2. The Slow Trade movement in Malaga 2121 involves the manufacture and distribution of goods from the city in a way that is inexpensive, labor intensive, and craft based. All of these are Slow, but they employ many people in enjoyable jobs and the end products are of good quality and long lasting. Most Malaga goods are traded around Spain, but some are exported on sailing ships to preferred Mediterranean trading ports such as Antalya, Athens, and Lazika.

3. The Slow Work movement in Malaga 2121 offers Malagans the opportunity to work a three-day week, enabling them to devote more time to leisure, family, their community, or maybe to growing and cooking their own Slow Food. There's maybe less income per person to begin with but more jobs to go around. This means that there is a greater level of social equality coupled with fewer social problems, factors leading to an overall increase in the quality of life.

Malé 2121 ¤ An Eco-Island Capital

The city of Malé, the capital of the Maldives, has more than one hundred thousand people occupying a one-square-mile island. It's the smallest capital city in the world in terms of area and also one of the most densely populated. The coral island that Malé rests upon reaches a maximum height of eight feet above the sea. Although, if you count the garbage dump on the island, this height more than doubles.

The diminutive presence of Malé renders it extremely vulnerable to changes in environment, such as sea-level rise and the increasing acidity of the ocean—both of which erode the coral that makes up the island. To contend with this destruction of a base to build upon, Malé 2121 adjusts to grow upward and just a little bit outward over the sea in order to both increase living space and raise the infrastructure high enough above any projected sea-level rise. At times, this will involve imaginative construction techniques that might relinquish lower levels of buildings to the rising tides. The slow decay of these buildings will also add alkalinity to the coastal sea, counteracting the coral-corroding effect of the sea's increasing acidity.

Another problem for present-day Malé is the shortage of fresh water, exacerbated by the increasing number of tourists arriving every year. Today, when the water supply suddenly dries up because of disaster or mismanagement, planes and ships from India are quickly sent at vast expense to make up the shortfall. In the future, though, India's response may not be so accommodating, especially if the Maldives start voicing concern about India's lackluster performance against global warming.

Therefore, in Malé 2121, water conservation and water recycling become imperative, with strict water quotas per person, to the point where shower minutes are rationed and baths and pools are banned. Throughout the city, and around its perimeter, there are extensive ecology zones where wastewater can be processed through plant-based recycling systems. An engineered wastewater factory, or a desalination facility, is likely to take up space and repulse tourists, but one highlighting the wetlands of the Indian Ocean makes the capital more attractive.

Mexico City 2121 ¤ Rebirth of an Aztec Lake

Mexico City is a place of environmental infamy. Smoggy air shrouds drab tower blocks and rusty factories, pierced only by the occasional glassy skyscraper or cathedral spire. In Mexico's capital, extreme wealth and ruinous poverty are intertwined with four million cars and fifty thousand unregulated businesses in a chaotic orchestra of traffic, dirt, and noise. The city sits in a basin surrounded by high mountains, a geographical bowl that traps the motionless smog and polluted water within it every day. Regular flooding afflicts the city when rainwater runs from the surrounding deforested mountains into the concrete jungle. The flooding gets worse year by year, since the overextraction of underground water is leading to the subsidence of the city, sinking it farther below the flood line. Some areas have sunk as much as thirty feet since the city was founded.

The city rests, for the most part, on an extinct lake, Lake Texcoco, which was drained, rechanneled, and filled in during the seventeenth century, then paved over in the nineteenth and twentieth centuries. By 2121, however, Lake Texcoco has been slowly rehabilitated to its former glory, providing for multiple economic and ecological benefits. First, it slows the flow of water to a manageable speed, and it soaks up the excess water to stop flooding. Second, it provides an ecological base from which to nurse forest trees and animals that can recolonize the surrounding hills and mountains, which together act to regulate water runoff as well as mitigate the risk of erosion and landslides. The lake also provides for wet and dry community gardening, whereby areas are set aside for low-income families to work the land to supplement their food needs. The lake also acts as a partial wastewater zone, filtering and recycling the wastewater of some of Mexico City's twenty million residents. Besides this, the resurgent lake provides a recreational area for Mexicans to get back to nature right within the heart of their city.

Minsk 2121 ¤ The Furry Roof Movement

Minsk is the capital and largest city of Belarus. The winters can be terribly cold, staying below the freezing point for many months. Temperate forests and gentle hills surround much of the city, all verdant and lush in the summer but covered in snow during the winter.

For seventy years, Belarus was forced to be a part of the Soviet empire, and since becoming independent in the 1990s the country has often been pressured into dialogue with Russian authorities about a possible reunion. By the mid-twenty-first century, the pressure may well have become too much. The need for Minsk to rid itself of crippling financial debt and the desire to secure cheap Russian gas will probably have brought Belarus back into the Russian Federation. A strong desire for independence is likely to linger, however, and in the early twenty-second century it will be manifest in a strange architectural form. To be less reliant on Russian gas for heating their homes and offices, nationalist Belarusians will cover their buildings with a furry insulation material that mimics the fur of the local variety of brown bear.

This Furry Roof Movement will grow to become a signal of resistance to Russian control. Soon, everybody with a gripe against Russian authority in the whole of Eastern Europe and Eurasia will be insulating their roofs in this manner, as an act of defiance against the gas-powered Russian hegemony. At the same time, they'll be striking a blow against the release of greenhouse gases. One day, sometime around 2121, utopia will come to Minsk as greater Russia breaks up once again, allowing Belarus to regain its independence.

Moscow 2121 ⌗ To the Future with Aerogaz

When the Americans decided to build a moon rocket, the Apollo/Saturn, to carry astronauts to the lunar surface, the Russians began to build one too, calling it the N1-L1. The Apollo/Saturn landed Neil Armstrong on the moon in July 1969, but Russia's gigantic N1-L1 exploded spectacularly in the Soviet desert air as soon as it was launched.

When the French and British started up a supersonic jetliner, the Concorde, the Russians decided to build one too, the Tupolov 300, nicknamed *Concordski*. The British Concorde by itself made fifty thousand flights during a thirty-five-year period. However, the first Tupolov 300 crashed in spectacular fashion at the Paris air show in 1973. An improved version later crashed during delivery to its airline company.

When the United States started building a space shuttle in the 1970s, the Russians began planning one, too. The US shuttle program completed 135 missions from 1981 to 2011, including the launching of the Hubble Space Telescope in 1990. The Russian space shuttle, named *Buran*, was used once on a remote-controlled flight in 1988. Fortunately, it didn't crash or explode, but it was mothballed straightaway to a storage hangar just outside Moscow.

If Russia's mimicking of the West's aerospace projects keeps resulting in gigantic mess-ups, perhaps we should fear the reestablishment of an airship fleet in Western Europe, in case Russia copies the idea and some mighty *Hindenburg*-type explosion is the upshot. However, given the simplicity of the design and the abundance of Russian gas, the Russians may in fact succeed in building, by 2121, a fleet of transcontinental airships to replace their expensive and unsafe jetliners. They are quite slow, but most Russians are used to taking days to cross their nation anyhow, on overnight trains like the Trans-Siberian Express, so an airship that takes the same amount of time might still be commercially viable.

How might Moscow 2121 be classified as *ecotopian*? First, unlike other megacities around the world, Moscow 2121's residents don't have to fight over where to put new airports and where to locate noisy new flight paths. This process pits neighborhoods against each other and tends to expand cities ever outward. Second, by ridding itself of the traditional aviation industry, Moscow could save the carbon credits that jetliners would use and trade them to invest in other eco-projects in the city—cleaning up the Moscow River, for example.

Mountain View 2121 ¤ The Search for ET

Mountain View, California, is the home of Google's headquarters as well as the SETI Institute, the former devoted to colonizing cyberspace and the latter devoted to the search for extraterrestrial intelligence in outer space.

Members of both organizations seem to be of the opinion that once we contact hyperintelligent aliens they will be able to teach us new and wonderful things about science and technology. They also believe our contact with aliens is imminent, if not within the next few decades then certainly by the next century. On the basis of this belief, we arrive at Mountain View 2121, where we find twenty-second-century biologist Dr. Sidney Schnoll, who has earned many prestigious awards by sticking things into starfish and hurting them. The starfish he stuck things into all wriggled in pain, and many thought that Dr. Schnoll would have been aware of this, but, alas, no. His mind was on the science.

Dr. Schnoll is also a keen SETI-ist, so he spends much of his spare time energetically broadcasting messages into the heavens in order to contact intelligent space aliens. Schnoll's trust in the essential goodness of all science and also in the universal benevolence of technology—alien technology included—led him to believe that contact with any technologically advanced alien would immediately make the Earth a far better place, since it would provide humanity with beneficial knowledge and super-duper useful gadgetry.

The dubiety of this assumption dawns upon Dr. Schnoll only when, after a night of particularly enthusiastic broadcasting, a hyperadvanced alien patrol ship lands near his house in Mountain View before sending out powerful probes that were pushed painfully into his head. He wriggled madly in pain but to no avail.

All was not a total loss, however, since the sad grunts and screams that he let out when each probe was wiggled around in his head were imaginatively used by the aliens to communicate benevolently with the California starfish.

Moynaq 2121 ⌗ The Port with No Sea

Ecotopia 2121

¤ ¤ ¤

In the 1950s, Moynaq, in Uzbekistan, was on the shores of the Aral Sea, once the fourth-largest lake in the world. Since then, the Aral Sea has been dwindling in size due to the overdrawing of water from the rivers feeding into it. This water has mostly been used to irrigate the huge cotton farms of both Russia and Uzbekistan. The Aral Sea has shrunk so much during the past fifty years that it has become a series of near lifeless, polluted, slithery lakes. In the 1950s and 1960s, Moynaq was a major seaport and fishing town. Now it is stranded some forty miles from the Aral Sea. The city's shipyard is completely dry and dotted with rusting ships.

Meanwhile, the new land that has emerged from the receding lake has been renamed Aralkum, the Aral Desert. Every now and again, a violent dust storm whips up the sands of the Aralkum, blowing them westward across the rest of Eurasia. As if this is not bad enough, the sand is laced with leftover chemicals from toxic pesticides.

As the Aral Sea receded, Moynaq's economy was destroyed, and now the remaining residents must eke out an existence however they can from their new desert. A small ecotourism industry has sprung up as people from around the world travel to the town to gaze upon one of the world's greatest eco-tragedies and to view Moynaq's haunting ship graveyard. Many local people still hope that, one day, the Aral's rivers will flow once again and the lake will grow to be alive once more. By 2121, this has not happened, but the residents creatively reconstruct their town to include a series of buildings celebrating their nationally renowned teas and symbolizing the seafaring past of the city.

Mumbai 2121 ¤ The Rise of Itopia

Mumbai is the largest city in India, possibly the largest in the entire world. Mumbai's densely packed, overflowing, chaotic form means the city authorities cannot really keep tabs on how many people there are within it, but some say twenty-five million is a good guess—with an extra thousand arriving every day by rail, on wheels, or by foot. To say Mumbai is a city of contrasts is an understatement. Mumbai has an estimated one thousand billionaires living within it as well as an estimated seven million people living in slums and some half million people living in slavery. In some areas, glamorous high-rise towers dominate the scenery, while other areas comprise squalid shantytowns.

Fast-forward to the future: Mumbai 2121 is a designated zone of infotech excellence where IT is cultivated, researched, developed, and then rolled out across each town and city in the nation. This *city within a city* will go by many monikers, none of them very original: "the City of the Future," "India's Smartest City," or, as Mumbai's twenty-first-century infotech class like to call it, "Itopia."

To build Itopia, India's infotech corporations work together with Mumbai's real estate moguls and construction companies to clear away slums, drain wetlands, and build their ten-square-mile infotech paradise. It is then populated with ambitious IT professionals who get to live and work in a clean and green tree-lined setting and who commute on uncrowded streets in futuristic, self-driven, solar-powered cars. Really, Mumbai 2121 is not much more than a pristine gated community for techie-nerds, but like other techno-vanity megaprojects around the nation, Itopia will be promoted as being for *all* India. Of course, Itopia will need some I-security features to keep out the poor of Mumbai so the roads and infrastructure don't get all jammed up and Itopians can work productively in peace and tranquility.

Building Itopia won't just happen because of a few ideas put forward by visionaries, though. It will be a massively expensive proposition for both public bodies and the private companies involved. There will be generous financial incentives offered to private companies to invest in Itopia, including tax breaks, free state-built infrastructure, and favorable land laws. Apart from all the new inventions and commerce, another rationale for the Indian state to offer support for Itopia will be that it is Green. Not just any kind of Green, mind you; it will be *smart Green*. For example:

- *Smart windows* will open and close automatically depending on the prevailing weather conditions. If it's too hot outside, the windows will close. If it's breezy outside, they will open. If the sun is shining brightly, the windows will detect this and become more opaque to keep the Itopians nice and cool inside.

- *Smart fridges* connected to smart cookers will burn just the right amount of gas to make Itopians their favorite dishes just the way they like them, though maybe with a bit more Vitamin B added if their smart toothbrushes detect that they were deficient that morning.

- *Smart trash systems* will suck domestic waste through underground chutes, where it will be automatically sorted and recycled, buried, or burned for fuel. These chutes will be connected to all apartment buildings and offices. Consequently, there will be no untidy street-corner trash cans or noisy garbage trucks in Itopia.

- *Smart toilets* in Itopia will know just what you excreted, and they will use the exact right amount of water, not a drop more, to flush it away.

- *Smart sewers* will ensure that wastewater in Itopia will hurtle through pipes to recycling plants at a hundred miles per hour, faster than the average Indian train. The pipes will filter what passes through them to divert recyclable elements back to a food production system. Initially, the smart sewers will also send nitrates and phosphates off to fertilize pretty Itopian gardens; however, over time the techies will decide they're not that interested in gardens, so they'll start building even more smart high-tech buildings on top of them.

- *Smart eco-security* will monitor potential environmental transgressions and alert Mumbai 2121's "Green cops" to take care of the transgressors.

- *Smart traffic lights* will monitor the city's self-driving cars and send them along the ideal route.

- *Smart parking* will deploy sensors around the city to monitor when spaces open up and guide cars to the best spots along the most fuel-efficient routes.

All of this will be very utopian for those living

within Itopia, but for those outside, it's another story, because:

- Many Mumbai slum dwellers will not have any kind of toilet, let alone a smart one. They will often be forced to defecate into a bag and then fling it into an alley—a so-called *flying toilet.*

- Most Mumbai residents won't have a car, let alone a self-driven one. They will move around the city on foot or by bicycle, both far more eco-friendly than smart cars; nevertheless, they are far more dangerous because of all the motorized traffic they have to contend with.

- Many Mumbai dwellers won't have regular access to electricity, or plumbing, or a sewage system, or education or training. And for sure, none of these people could ever afford to live in Itopia. Once upon a time, they might have gotten unskilled jobs doing cleaning or recycling for an infotech company, but in 2121 Itopia will be so *smart* that even these jobs will be taken over by IT robots.

- When the Indian government decides to build any kind of new urban techno-project, it usually involves kicking the poor out of some place and knocking down their houses. This will be the fate of those who have to make way for Itopia.

All in all, Indian society is full of various social barriers between the haves and the have-nots, and Itopia will make for yet another. One good thing, though: this reliance on smart technology will eventually make Itopians so incapable of thinking for themselves, so unused to making decisions, that they might easily be out-thought and overpowered by the millions of slum dwellers and slave workers in Mumbai.

Nador 2121 ⌑ Oilgae and Eco-punishment

Nador is a port city of nearly two hundred thousand people on the Mar Chica lagoon of the Mediterranean coast of North Africa. In times past it was a Berber city, a Phoenician city, a Roman city, an Arab city, and a Spanish city; now it is a Moroccan city.

Nador is currently going through an economic boom via two industries: metals processing and tourism. They seem to be working against each other, though, since the cultural and natural beauty is being eroded by a polluted lagoon and polluted air. Occasionally there are spills and run-off from the metals industry, which some believe are dangerous to the health of local residents.

One problem that Nador confronts every year is the algal blooms in the lagoon, which mess up the quality of the water and create a nuisance for the tourists. In Nador 2121, however, these annual blooms are transformed from a problem into an opportunity, as the algae is extracted and used as "oilgae" in bioreactors to create energy.

To control the metals industry, a different approach is taken, which comes about after a massive industrial accident sometime in the early twenty-second century, when thousands of tons of chemicals leak from a metals factory into nearby schools and communities, and eventually into the lagoon. The leak leads to the death or incapacitation of dozens of children due to heavy-metal poisoning. The factory owners get away with it—much to the disappointment of the whole city—since Moroccan law allows the company to be charged for environmental crimes and punished with fines but individual bosses cannot be held accountable.

In this scenario for Nador 2121, the bosses are made culpable for any life-threatening environmental pollution their firm produces—and the punishment is public execution. The bodies of the convicted managers are then reprocessed, like the algae, into biofuels at special public events. Through these measures, the importance of a clean and Green environment is conveyed to all Nador's industrial leaders.

Namibe 2121 ¤ The Silicon Desert Town

Namibe is a coastal city in Angola, adjacent to the Namibian desert, from which the city takes its name. Namibe's main industry revolves around its port, where fishing vessels dock along with container ships that carry away exports from around the country.

Today Angola is enjoying a mineral boom—its economy is rapidly growing via mining, though with great disparities of wealth also surfacing and with many environmental problems, too. Presently, the Angola government is seeking to develop the Namibe desert, and one of the minerals it is prospecting for is uranium. Uranium mining is a dirty business. Wherever they have been sunk, uranium mines always pollute waterways, contaminate communities, and displace people, and they usually contaminate animals and plants with radioactivity. Often, the mines never become profitable, since the market is highly unreliable and the cost of environmental cleanup is enormous.

Still, uranium mining promises a flow of cash, usually by way of government investment, and someone is pushing for more mines in order to get their hands on this cash, even though the communities where the mines are located always protest against them. Usually, governments are led to believe they can upscale their profits if they process and enrich the uranium to make it more valuable. Thus, we can imagine sometime in the future of Namibe a huge mobile factory that churns through rock and sand for uranium ore, then processes it into uranium oxide and purifies it into uranium metal, ready for export. All this value-adding would, of course, involve heavy carbon dioxide emissions, great usage of water resources, and an expanded risk from more radioactivity. Even if the final product sells at a higher price on the market, any form of international economic turbulence might plunge the industry into a huge financial loss for Angola.

Fortunately, in 2121 the Angolan finance minister suddenly recognizes the economic nonsense that is the uranium industry. Just before the project is green-lighted, she approves a transformation of the industry to mine silicon for solar panels. And as a further value-add, the mobile factory turns the silicon into complete and finished all-weather solar cells, ready to ship around the world.

New Orenburg 2121 ⌖ Ecotopia on the Rocks

Perhaps the Peak Oil idea—that we are doomed to run out of fossil fuels before long—is so completely mistaken that by 2121 it might be but a quaint, faded memory. By then, the oil and gas fields of the Arctic and Antarctic will have been opened up. This will be made commercially feasible due to the warming of these areas and the subsequent melting of the ice sheets—making the development of oil and gas reserves much easier.

At present, the Antarctic continent is protected by a two-mile-thick ice sheet and also by the Antarctic Treaty, which forbids oil exploration. By late this century, though, the ice sheets will have greatly subsided, opening up the Antarctic lands and seas to the possibility of resource exploration. The Antarctic Treaty is also scheduled to lapse in the 2040s—perhaps foreshadowing an Antarctic land grab.

And so presented here is a new city in Antarctica, New Orenburg, perched on Antarctic rocks exposed by the withdrawn ice cap. Initially set up by a Russian gas company, the town was supposed to provide a homey and steady environment on the ice sheets for workers and their families, but it quickly becomes just another industrialized gas-extraction facility perched in the Southern Ocean.

The name references Orenburg, a frontier town set up in the Ural Mountains in the late eighteenth century by the Russian empire as it pushed to conquer the Asian continent. Today, Orenburg in Russia has a population of half a million and calls itself "The Gateway to the East." New Orenburg 2121 will have a population of maybe ten thousand and will call itself "The Gateway to the South."

Some might see New Orenburg as a dystopian vision for Antarctica's future, since it risks the onset of eco-degradation, wildlife extinction, and huge oil slicks or gas explosions. But it has at least one claim to eco-friendliness: instead of using pipelines that would crisscross the continent, disrupting the landscape, New Orenburg uses giant container balloons, which fill up with natural gas and are dispatched to float safely and serenely to markets around the world.

New York 2121 ⌘ Revolutionary New Learning

Date: May 1, 2121. Location: the main campus of New York University, following this year's commencement ceremony. Long after the crowd had dispersed, half a dozen new graduates were sitting by the fountain in Washington Square Park. They were looking at the bandages wrapped around their arms. Underneath the bandages, their wounds bulged and bubbled with agitated blood cells, as though their new knowledge was not yet quite compatible with their bodies. Although intravenous education shepherded students through the drudgery of undergraduate courses a lot more quickly than the standard four years, it did place enormous strain upon the brains and bodies of some recipients. These remaining graduates were feeling a tad nauseated.

Despite this, they bunched together, scrolling for the job-board vacancies on their smart devices. Newly graduated, they now had to find work. Nobody seemed to be in a hiring mood. They did notice, however, that the university president's speech was printed in full on the NYU website:

You see before you a proud man. Of what am I proud, you ask? Of you! I am proud of your computational skills, I am proud of your techni-cal prowess. And, of course, I am proud of your loyalty to the ideals of this university. We used to be a tiny liberal arts college, with a quaint little library and an art gallery. But it is technology that changes worlds. The nanobot-filled fluid that flowed into your arms for the past six weeks is a revolutionary technology. No more boring lectures. No more boring libraries and art galleries. And with your help, we can prospect for more and more rare metals far and wide to make more and more nanobots and then work to spread this technological revolution throughout the world!

The graduates did feel a tinge *revolutionary* but not in the way meant by the president. Sitting, nauseated, under the clear blue New York sky, they felt rather cheated by the techno-learning program. And they also felt indignant that their university was planning to colonize the world with it, probably smashing its way into pristine landscapes in search of more rare metals. However, with no job, no money, and only each other to consult, they felt powerless to stop it all.

As they sat nursing their nausea, grimly resenting how the university had seduced them away from traditional learning, a new idea suddenly entered their discussion. "Maybe we could use the nanobots for good. Maybe we could reprogram them for a *social* revolution. And maybe we can even reprogram them to spread around the city like pollen in the wind or like a contagious disease."

Nanobots are nano-sized robots used in various medical treatments. By the twenty-second century, they are imbibed by patients to attack all manner of illnesses, including those in the brain. The university simply rejiggered this medical application into a learning mode and pumped the nanobots into the bodies of their hapless students.

"Couldn't we just reprogram the nanobots with new ideas?" asked one graduate. "Something that would act as an antidote to university power?"

"Like what?" asked another student.

"I don't know. Marx? Gandhi?"

"And the eco-philosophy of the Neo-Luther Robo-Buddhists?"

The other graduates laughed a little, then shrugged their shoulders and nodded their heads. One of them raised a toast to the idea with a bottle of green tea in her hand, pledging their allegiance to resisting intravenous learning everywhere. She loudly called for the sacking of the president's office. "Revolution here we come!"

she yelled, eliciting a rousing cheer from the others. "Careful, you're going to spill the tea," one of them warned.

The new graduates hatched a plan to dose a dozen NYU classes with these reprogrammed nanobots and then let them spread like an epidemic among the general population. On one crazy green tea–powered night, they managed to steal some nanobots from one NYU laboratory and plug them into their own computers to reprogram them. Then they ran around campus all night injecting them into various intravenous learning systems.

◻ ◻ ◻

For breakfast, the students retired to a student bar near Washington Square to take coffee shots and drink Bailey's shakes while they told the other bar patrons about their wild night. For hours they spoke passionately about making the world a better place, how they reprogrammed the nanobots, including a special subroutine to build a new library that would float serenely on the rising sea level along New York's coastline. When they slowed down a bit, they bought

New York 2121

◻ ◻ ◻

more drinks, and then continued recounting the details of the story to an ever-growing audience in the bar.

By late that afternoon, they'd all slouched off to a quiet corner to sleep, confident that a new batch of nanobot-charged graduates would soon rise up and revolt against the president and his imperialistic NYU.

Except . . . the graduates had made a small mistake. Instead of plugging their reprogrammed revolutionary nanobots into the pump that led to the students, they had plugged them into a feed that led to a broken-down cola vending machine. Their revolutionary nanobots were safely secured in the vending machine, and they had all probably dissolved away into nothingness by the morning.

The amazing thing, though, is that a revolution did occur in New York.

Some weeks later, NYU was ransacked by a large mob of students, and the intravenous nanobot learning system was plundered and destroyed. After they had occupied the campus for a few festive days, the president was forced to resign. New leaders were elected who were committed to "real learning, not machine learning," and who immediately abandoned NYU's devastating rare metal mining projects around the world.

As restitution for the university's past ecological misdeeds, an outreach program was started, in which environmental education was provided free to all New Yorkers. This program transformed New York City into a very Green city. Within a few years, seagulls could fly in clean air above a rehabilitated harbor teeming with healthy, tasty fish.

But—you may very well ask—how could such a revolution happen when all that the graduates had managed to do was pump their reprogrammed nanobots into a broken-down vending machine?

Well, in that one day of fervent storytelling in the Washington Square bar, they had so inspired their audience with hope and passion that all who heard the story rose up to rally an even larger idealistic group of student activists. This group kept snowballing in size, and eventually its members headed off to occupy NYU in a merry, unstoppable protest.

It was not technology that changed the world but enthusiastic storytelling!

Ecotopia 2121

¤ ¤ ¤

Nizhni Novgorod 2121 ⌖ City of Self-Sufficiency

In the thirteenth century, the city of Nizhni Novgorod cleverly escaped a Mongol invasion by being too insignificant to attract the Mongol army's attention. In the ensuing centuries, however, it grew to become the third center of importance in Russian life, after Moscow and Saint Petersburg. Nizhni Novgorod served as a banking capital for the Russian empire and was a leading military and industrial city as well. It was here that the Stroganov family set up their business empire.

Then, during Soviet times, it was pronounced a "closed city." That is, it was closed off to foreigners and Russian citizens who had no business there. The main reason for this was to protect its array of military facilities.

Nizhni Novgorod was the hometown of Russian literary hero Maxim Gorky, who wrote about the dismal life of the proletariat in the city prior to the 1918 socialist revolution. When he died in 1936, the Soviets renamed the city Gorky in his honor. In 1992, by popular demand, it reverted to its original name.

In the 1920s, Henry Ford visited the city to instruct the Soviets on how to set up their huge GAZ (Gorky Automobile Zavod) factory, which is still the biggest factory in the country today. The commercial success of the GAZ-made cars has waxed and waned throughout its history, no more so than nowadays, when year by year the fortunes of the company swing wildly with the world's changing economic situation and Russia's status on the fractious international trade scene.

By 2121, the constant unpredictability of car sales forces GAZ to reconsider its main product line. Working with the local government, the city decides to expand its network of streetcars, for they are always very well patronized and can help Nizhni Novgorod grow a little Greener by cutting down on air pollution. Traditionally, for decades, the city's streetcars had been supplied by Skoda, a Czech company, but in 2121 they are all made right here at the GAZ factory and specially customized to the local setting.

Nuuk 2121 ¤ The Green Grass of Greenland

Greenland was given its name by a Viking, Erik the Red, not because it was green but in an attempt to attract colonists from Scandinavia. According to the sagas retold to this day, his propaganda was hugely successful, since he convinced hundreds of Icelanders to uproot and resettle in Greenland. They ended up founding a settlement very near to where the current capital of Greenland, Nuuk, is located. Much to their dismay, Greenland wasn't green; it was even whiter with ice than Iceland.

Under global warming, though, Nuuk is slowly transforming from icy white to a vibrant green as more grasses and shrubs take hold and spread in warming environs. Sometime in the next decades, the first naturally seeded ash tree will likely emerge into the crisp Nuuk air. When this first ash tree grows, it will be heralded by Greenlanders of Viking descent as the resurrection of *Yggdrasill*, the ancient, mythical tree that connects and unites the various dimensions of the world. By 2121, this tree will be honored in the layout of the town as the centerpiece of the city. It will symbolize the rise of Greenness in Greenland, honoring nature and representing the connection between past and future, between Greenland and the promises of Eric the Red.

For Greenlanders of Inuit descent, however, the tree will likely provoke mixed feelings. They may see it as an invasive species, symptomatic of European colonization. Or they may see it as a sign of Earthly beauty, of warmth, growth, and opportunity.

For all Greenlanders, Nuuk 2121 will be very different from Nuuk of today. In 2121, they at least will be able to grow their own crops, making them less dependent on imports from Europe and North America. And as the warm season gets longer, more tourists will seek out a view of the otherworldly landscape. By the same token, however, it's likely that the landscape will rapidly be despoiled and pockmarked as Greenland's mineral riches are exposed from under its melting ice caps and mining rises to be the primary industry. These changes might bode well for invasive flora, but they could also spell doom for Greenland's fauna, especially the sea life. The mighty Greenland whale, for instance, which has the largest mouth of any creature in the entire world, might have nothing to eat in the future, as the plankton die out in a sea that has become too warm, too acidic, and too polluted.

Ordos City 2121 ⌻ Walled City in a New Desert

The Ordos are a Mongol people who lived for many centuries on the Ordos plain in what is now northern China. At present, they number about a hundred thousand people spread over three Chinese provinces. For centuries, the Ordos plain was a prairieland, with rivers and streams in abundance, making it a vital pasture zone for Ordos farmers. However, over the course of just one generation it has become a desert. When the Chinese took control of the area in earnest in the late twentieth century and subjected it to a combination of devegetation, hydrological overextraction, and megasized coal mining, the prairie became denuded, dried up, and desolate.

The push to start a massive coal industry within the Ordos plain worked out very well commercially, and it meant money could be spent on the rapid building of a gigantic new city. Ordos City, built in the last decade from scratch, now stands complete with new tower blocks, broad avenues, malls, and luxury apartments. The only thing the city lacks is people.

Ordos City was built to house one million, but currently only about ten thousand workers live there. The government says, "Just wait—people will come," and in the meantime officials are busy convincing the Ordos people to move from their farms and villages, ravaged by the growing desert, into the new city. However, the rent is far too expensive for farmers and villagers, and there is nowhere to grow food, so most of the Ordos stay put. Nevertheless, new electricity plants are installed, new highways are constructed, and the city grows ever larger, though without people.

As the decades of this century pass, the city will likely be abandoned by China's government when all the coal has been ripped completely from the land. They won't make any official announcement. That would be too embarrassing. But the surrounding rural people can see it happening, as Ordos City becomes a ghost town over the course of one long winter night. When day breaks, the Ordos people ponder the situation for a few minutes before they move in to occupy the empty city. By 2121, not only have they made the abandoned city their home, they have recycled surplus infrastructure to build a great wall around it. They tell outsiders that it's to keep the desert at bay, but everybody knows it's really to keep the Chinese government out.

Oxford 2121 ⌀ Tranquil Streets of a Retro-future City

Oxford, England, 1821. If you ambled gently along the banks of the River Thames, then across it, past Christ Church Meadow, and through New College Lane just after sunrise on a day in May of this year, you had a good chance to spy dreamy white swans, languid willows, and luscious fresh fruit taken from the fields.

Oxford, England, 1921. If you ambled once again along the banks of the river and then across the meadow and through New College Lane on a day in May of this year, the swans would be a little harder to find. The noise from all those dozens of loud new automobiles would be scaring them right away.

Oxford, England, 2021. All the roads around town are so jam-packed with cars every single day of every single month in this year that it takes nearly ninety minutes to travel just a mile. So dump the car and take that slow amble along the river—it's got to be quicker.

Oxford, England, 2121. The swans are back, dreamily floating again, each morning and every day of May, and New College Lane is peaceful and enchanting once more, for Oxford has banned cars. In a few hours the entirety of the city's scholars and townsfolk will be awake and ambling.

What's so bad about cars that Oxford 2121 decides to ban them altogether? Well, for starters:

1. Carmaking leaves a giant ecological footprint all over the globe, especially with all the steel, rubber, plastic, paint, and glass that's needed to produce a single car.
2. During its lifetime, even the cleanest car will emit toxic chemicals and climate-changing gases.
3. When a car dies, it leaves a decaying, rusting body that takes up land space and slowly pollutes the environment.
4. Fueling a car often involves the need to wage violent international interventions and incur further environmental despoliation.
5. Car accidents are a leading cause of death among young people, in England and around the globe.
6. Exhaust gases from cars cause respiratory diseases worldwide, often paid for disproportionately in health terms by non–car

Ecotopia 2121

¤ ¤ ¤

users (including children and the elderly) and in financial terms by the public purse. Cars also have a big role in the global cancer epidemic; about half the cancers attributed to outdoor pollution result from vehicular exhausts.

7. The roadways needed to facilitate car use have (a) destroyed many natural environments and communities, (b) caused the incursion of urban sprawl into nature and the countryside, and (c) ushered in further dependence on car ownership as the only possible transportation option.

8. Cars have encouraged (a) conspicuous over-consumption, (b) social isolation (among both those who own them and those who don't), and (c) increased time and distance between places of work and places of living, giving rise to further problems associated with long-distance commuting.

9. Cars punish non–car users by encouraging the elimination of sidewalks in suburbs and pedestrian plazas in city centers, and they · make for a noisy, stinky, stressful, and dangerous environment for those who walk, especially for children and their minders,

and for those with limited mobility like infants, the elderly, and people with disabilities.

10. Cars may be seductive to look at individually (if you don't mind the noise and stink they make as they whizz on by), but traveling bumper to bumper all together, or parked all over the city in great, desolate concrete wastelands, they tend to turn pretty and peaceful areas into noisy, ugly, crowded places.

Despite the allure for car fans and their glorification by the car industry, these "machines of death" are not universally loved. This is especially the case among those who cannot afford or cannot operate cars. Cars create a strongly divided society. Those people with cars in a car-dependent city exert great physical power on a daily basis over those who do not have a car—threatening them with injury, risking their safety, cutting off their options to walk or cycle freely, and poisoning the air they breathe. As one example of the socially divisive effects of cars, consider that the highest concentration of harmful gases caused by motorized traffic usually is found at the height of three

Oxford 2121

◻ ◻ ◻

feet above ground level. Hence, in most cities of the world, children breathe more toxic air than adults and their health is jeopardized even more strongly than that of adults.

But to ban cars from an entire city? All the way to the edge? It's not as far-fetched a scenario as it may seem. Even today, in the early twenty-first century, parts of Oxford city center are car-free pedestrian zones, and many cities across Europe have much more extensive car-free zones. By 2121, it might not be impossible for an entire city to ban cars right to their outer limits.

It's appropriate to admit here that one person's vision of utopia may well be another's vision of hell. If you are loathe to lose your car (if you are laboring under the misconception that it provides you with security, convenience, and freedom, or you just absolutely need it to get to work or take the kids to school), it is likely that Oxford 2121 will not be of much inspiration to you. This attitude is, for sure, widespread across the world, supported and bolstered within city economies and the car-glorifying media. However, rest assured that an evolving ecotopian city will offer citizens a chance to pursue other forms of freedom beyond the superficial freedom of mobility that a car offers. A utopian city—as it is brought into being or evolves over time—acts to encourage each and every person to learn and adapt to the new or the evolving utopian form. Citizens can thus teach themselves (as individuals, as communities, and as societies) to be happy—even happier—without cars.

Most people in the free world believe that there are only two real certainties: death and taxes. In Palo Alto 2121, located in California's Silicon Valley, they laugh in the face of both. Here, they get tax breaks from the California state government to pursue weird techno-dreams that defy death. One dream is to convert all the thoughts and experiences within a person's mind into digital form and then upload them into some form of alternative biomachine—an avatar, if you like—that can go on living forever, even when their own bodies and brains turn to dust.

On the streets of Palo Alto 2121, the first mass-produced avatar is now a common sight. Here, the three-legged, one-eyed avatar is fashioned from dead physical remains of digitally uploaded humans along with lab-cultured mammalian body parts. This three-legged, one-eyed variety is selected because it is cheap (for Silicon Valley millionaires, that is) and easy to make and a little more stable than two-legged varieties.

Some regard these avatars as kind of cute and colorful. Others see them as so abominable that they give them scary monikers such as "Frankenbeasts" and "Valley Zombies." Yet, for technophiles, they are a shot at immortality, an opportunity to live on in a postdeath world. However they might be viewed, the only cities that don't outlaw their free-roaming existence are all in Silicon Valley, and here they sell like hotcakes.

How is this anything like a utopia? Well, the avatars can serve to resurrect long-since-deceased techno-geniuses of the past. In Palo Alto 2121, an avatar of Nikola Tesla might bump into an avatar of Steve Jobs and, of course, the ubiquitous Einstein avatar, too. If these people were alive during the digital age, their brain waves and thoughts were converted into bits of information and uploaded to the avatar's computer. If they came from the predigital age, then some fragment of their DNA plus a digitized compendium of all their writings and recordings is uploaded.

According to avatar fans, these resurrected geniuses will go on to make the whole world a better place by creating innovative technologies, including eco-technologies and energy-saving devices. The avatars have an added "eco" advantage in that they have the capacity to instantly adjust their settings to maximize efficient use of energy and resources, unlike slow and stupid humans. They are also better able to survive whatever pollution is out there in 2121. Whatever toxic

Ecotopia 2121

¤ ¤ ¤

152

waste humanity can throw at the world might kill flesh-and-blood humans, but it won't hurt the hardy avatars. The Earth may get smogged out, the climate may be in chaos, but the avatars have robust bodies and fast minds to help pull them through.

Of course, like all new technologies, there are unforeseen risks. In the race to get the avatars on the market to serve dead and dying techno-geeks, and to make living entrepreneurs lots of money, most of the avatars are not perfectly engineered for the city environs. Sometimes, they clumsily bump into people, each other, and passing cars. Other times, they spew toxic effluvia in the faces of children or swing their third leg a little too eagerly and whack an elderly person in the butt. Many consider them so accident prone, annoying, and clumsy that most cities around the world end up outlawing them altogether for safety reasons. But in Palo Alto, they are free to roam the streets at will.

◻ ◻ ◻

By 2121, the city is swarming with these cyber-zombies. The living are outnumbered by the undead. A few hardy "lifers" stay on, but eventu-

ally, as the avatars become the majority and learn a thing or two about politics, they stage a takeover and start organizing Palo Alto for themselves. It might seem crazy that zombies could run for office, since they're not even really human, but their lawyers declare that the zombies are the artificial bodies of real persons manifested in a new form and that, as real persons, they deserve full human rights.

In the end, the avatars prove as clumsy at making city laws as they are at traversing the streets. To mitigate this, the avatars often invoke a rather unpredictable autoparalysis response to their own clumsiness, which freezes them in their tracks when they notice another zombie lumbering into a potentially dangerous situation—crossing the road, for example, or ducking under a tree. This autoresponse is never, ever programmed to the right degree of sensitivity or reaction, so some zombies just collapse into paralysis when they see another living being, and other zombies start spasming uncontrollably, flailing their limbs in all directions because the visual cues hitting their big eyeball cannot be processed effectively.

To deal with this, another autoresponse is programmed and rolled out, whereby the eyeball

Palo Alto 2121

◻ ◻ ◻

self-destructs if it is receiving too much information. In some zombies of a nervous disposition, the eyeball explodes after the ninety-ninth daily shutdown in order to quell disquieting inputs. This is particularly inconvenient when a zombie has already seen ninety-nine risky situations over the course of the day and has to head home across a bunch of busy intersections.

In order to remedy this situation, the Palo Alto city government erects thousands of signs all over the city, warning about potentially risky situations and the possibility of adverse automatic reactions. This in itself seems to do nothing but exacerbate the problem because it subjects the cyberzombies to further nervousness, prompting more and more of their eyeballs to explode.

Ecotopia 2121

□ □ □

Panama City 2121 ✕ The Rain Forest City

Panama City 2121 is flush with biophilia. This is not some strange tropical disease but a psychological affection for nature. Thus, the residents of Panama City 2121 love nature so much that they have immersed their city in vibrant wilderness. This provides them with both a free ecosystem and emotional well-being. In Panama City 2121, the forest of the Metropolitan National Park, nowadays located between the city and the Panama Canal, is allowed to recolonize the entire city, overtaking the concrete infrastructure that currently keeps it at bay. The architecture has become organic and biological, which is a style favored by locals because it is inexpensive and self-maintaining. It's also rather romantic and fun.

In the early twenty-first century, Panama City has a rather disturbing air pollution problem, but the Panama City air of 2121 is fresh, and freshly scented, due to the cleansing character of the forest canopy and the city's adoption of new ecological industries, such as Green horticulture, eco-medicine, and sustainable forestry.

Panama City's bays and canals are also rather polluted in the present day, their waters being sometimes unfit for human use. But Panama City 2121 has waterways refreshed by the forest's expanded natural filtering system. Children can swim in the canals around their homes, and they can also play safely on the shores of the Panama City bay.

The trees of the city forest provide electrical power for Panama City 2121. Like most living things, trees produce a small electric current, and this tree-made electricity flows wirelessly to microenergy devices that can work with the smallest of electric currents. This might seem like a wild idea, but even today there are research projects devoted to tapping tree electricity in order to power tiny remote sensor devices. By 2121, after decades of microenergy investment, the harnessing of such electricity may very well become more widespread.

Unfortunately, in the early twenty-first century, Panama's forests suffer ongoing destruction. In 1950, more than 70 percent of Panama was forested, whereas today it is less than 30 percent. Millions of acres of vegetation disappear every year. All this deforestation threatens the economic value of the country's most important income earner, the Panama Canal. The water in the canal, which keeps everything afloat and mobile, is slowly being drained away due to the declining capacity of the receding rain forests to produce silt-free water.

◻ ◻ ◻

So how do we get from the Panama City of today to the biophilic city of 2121? According to advocates of biophilia, there are five steps to follow:

1. Education about biophilia (in schools, in universities, and in outreach to the general public)

2. Practice and experimentation with biophilic techniques and products (by inventors, entrepreneurs, researchers, and do-it-yourselfers who wish to bring nature into their backyards and into experimental sites)

3. Expansion from the backyard (and experimental sites) into the schoolyard and into the courtyard of government buildings and corporate research-and-development sites

4. Commercialization of biophilia techniques and products (sometimes with government help and sometimes not)

5. A rise to common usage (again, sometimes with government assistance, sometimes without it)

So, in the beginning, somewhere in Panama City right now, enthusiastic urban experimentalists must work on their biophilic designs to perfect them and then work with teachers to educate children about the role and possibilities of biophilia. When the children grow up to become farmers, teachers, managers, scientists, and engineers, they are ready to press for investment in more research, more ecological instruction in schools, and more ecological practices in the public and private spheres. Within a few generations, people will be making money from biophilic designs, and new industries, including biophilic architecture and forest gardening, will stake their place within the economy and lifestyle of Panama City. Aided by the desire to preserve the Panama Canal, and with ecological awareness growing year by year, a biophilic land-use policy becomes the logical pathway for future urban planners.

This is a process of long-term social learning and technological change. One of the most important things to learn will be the idea that the economy can change from dependence on old-fashioned materials and processes to one utilizing organic materials and processes. So instead of concrete,

Panama City 2121

◻ ◻ ◻

plant fibers are used. Instead of petroleum, plant oils are used. And instead of relying on chemical combustion to produce energy, photosynthesis—capturing sunlight—becomes the main form of energy production.

One of the most important stakeholders in the development of biophilia in Panama City 2121 may be the shipping companies that use (and fund) the Panama Canal, since they have a vested interest in trying to preserve the canal's hydrological well-being. This means they may be very keen to support and invest in public-private partnership projects to reforest Panama City and convert the city economy into a Green one.

Ecotopia 2121

¤ ¤ ¤

Paris 2121 ¤ Space-Age Environmentalism

As an honors degree student in England—decades ago—I, the author, had the opportunity to do field-work for my honors project on the history of the Russian space program right there in Russia, visiting many Russian space facilities. At the time, the first British astronaut, a chemical engineer named Helen Sharman, had just been launched to the Russian space station *Mir*. To celebrate this, a couple of Russian and British universities set up a student exchange to share experiences of their respective space research.

One of the places I visited in Russia was the Tsiolkovsky State Museum in the small town of Kaluga. The museum was built during Soviet times to commemorate, and to publicize, the work of the Russian scientist Konstantin Tsiolkovsky. Tsiolkovsky was most famous for his diverse aerospace research in the years of the late nineteenth and early twentieth centuries. He was the first to design multistage rocket boosters, the first to design spacecraft airlock systems, and the first to design various kinds of aircraft, hovercraft, and jet craft. None of these were ever built by Tsiolkovsky, and all his grant applications to pursue research into them were rejected by the Russian Army and the Russian state when he submitted them, but that doesn't stop Russians nowadays from calling him the father of cosmonautics.

In his younger days, Tsiolkovsky marveled at the great new engineering structures of his age. In 1895, he was so inspired by the magnificence of the Eiffel Tower, then only just completed, that he designed a larger tower of similar form that would rise sixty miles, all the way to the edge of space. Within it, an elevator would move up and down to transport humans and cargo to orbit and back again. As yet, no one has proceeded to build such a tower, since no material is strong enough to support such a massive structure.

¤ ¤ ¤

As I progressed into my postgraduate studies, moving to Australia, I stayed interested in the space programs around the world, and I noted that many advocates of space travel justified it for environmental reasons. In Russia and in the United States, space fans said that humans need space exploration to find rare resources and to open up new territory for Earth's growing population. Many felt that space technology would help humanity develop innovative eco-technologies.

Most of them also talked about the way space-age photographs of the Earth ushered in a cultural revolution in environmental consciousness, citing the contemporaneous rise of the American environmental movement with the US space program, especially as astronauts sent back pictures of the beautiful blue Earth taken from far away in space.

Tsiolkovsky, in the early twentieth century, believed that colonizing space would lead to the perfection of the human race, as we transcend our Earthly home to spread to other planets and tap into new energies with superadvanced technologies. In Soviet Russia, the new communist state developed its own benevolent space expansion ideas via "cosmism," a mixture of communism and futurism, to encourage the working classes to adopt magnificent machinery that could propel the proletariat toward a social paradise here on Earth and maybe later in the rest of the solar system.

While fascinated with the Space Age and its dreams, I am not at all convinced it had any strong claims to being environmental. I interpret the environmental movement as a reaction against the industrialism and militarism associated with the space race of the 1960s, not as being inspired by it. I acknowledge that the photographs of the blue Earth from space are wondrous and beautiful, but I cannot help but be sympathetic to the thoughts of the philosopher Martin Heidegger, who said of them that they were like sweet good-byes to our planet, as though humanity were sighting the Earth in a rearview mirror as we sped away from it.

The space travelers themselves seem oblivious to this. For instance, while serving onboard the International Space Station in 2013, the astronaut Colonel Chris Hadfield sang the 1969 David Bowie song "Space Oddity" to an Earth-bound global audience of many millions. In the lyrics of the song, a spaceman called Major Tom sings a tale of his alienation from Earth and how he has to resort to drugs to deal with it. The song is something of a parody of Stanley Kubrick's spectacular movie *2001: A Space Odyssey,* which celebrated the monumental wonder of space travel. It was released in 1968, a year before Bowie released his song.

If the cultural contradictions of space travel are ignored by those who participate in it, then space travel's relationship to the arms race is also

Paris 2121

glossed over. You don't have to be a rocket scientist to realize that the launch vehicles that carried astronauts and cosmonauts into orbit were codesigned to act as missiles carrying nuclear warheads. The moonshots of the 1960s came about through Cold War competition, as part of a struggle for military techno-superiority; not as a way to work for the "common good" of humanity—or the environment.

Perhaps more scarily, in the 1960s American rocket scientists made plans to use hundreds of sequential nuclear bomb explosions to propel huge spaceships out into deep space, maybe toward other star systems. So far, these plans have yet to materialize, but there have been a series of small atomic-powered space probes sent to explore the outer planets, such as the Voyager and New Horizons missions, and the chief of the US space agency is also pondering the possibility of nuclear rockets to go to Mars. In space, maybe a little bit of nuclear material is not all that dangerous. However, the process of getting it *into* space is very risky. Nowadays, space rockets have disastrous malfunctions about one-sixth of the time. A large proportion of this one-sixth explode high in the atmosphere, and if an exploding rocket

includes a payload of nuclear material, the explosion would possibly vaporize the nuclear material and then scatter in a huge band around the globe. Eventually, the radiation would fall to Earth as rain and snow, dousing billions of people with radioactivity. Those living in the irradiated band could possibly inhale radioactive molecules into their lungs, resulting in cancers and lung disease, maybe at the scale of millions of people.

I published my concerns about space development in academic form and sent them off to an elderly but still very collegial Arthur C. Clarke, the well-known futurist and science fiction novelist (and the guy who cowrote *2001: A Space Odyssey*). He was kind enough to send me a nice reply saying how much he missed snorkeling in Australia, and he also sent me the preprints of his latest novel, circling in highlighter a paragraph about environmentalists who wanted to protect the pristine beauty of Jupiter from human interference. In some of these scholarly papers, I was hopeful that large-scale use of nuclear reactors and nuclear bombs in space would probably never eventuate because it was just too expensive to develop such projects. Maybe space agencies could blind the world with science about the supposed safety of using nuclear material in space, but if they wanted to launch a nuclear-powered rocket to Mars, someone would have to pay for it, and in all likelihood politicians would strike it down as being a big waste of money. However, Clarke told me of an optimistic little secret buzzing around in the heads of space fans. In a few decades, he suggested, Tsiolkovsky's vision of a space elevator might be realized, and this would rapidly bring down the cost of space travel (including the cost of getting huge nuclear-powered spaceships into orbit).

¤ ¤ ¤

So, taking the lead given to me by Arthur C. Clarke, I present Paris 2121: a prediction for the future of space travel. In this scenario, it is not the Americans or the Russians who work to make a space elevator but the French. France, after the United States and the USSR, was the third nation to reach space, launching its first satellite in 1965. Today, the French still have a strong space program, being a major participant in the International Space Station while also launching satellites to Earth orbit and space probes to other worlds.

By the early twenty-second century, after decades of research and development, the French space agency builds a space base in orbit and then unreels an elevator cable from it—which comes all the way down to the surface of the Earth. The cable is eighty miles in length, made of nanocarbon, and it descends from space to make contact with the Earth at France's space center in French Guiana in South America. The first elevator car starts running up and down the cable in 2121, ferrying both people and machines to and from orbit. The project is reported the world over as being "an engineering marvel."

As a financial enterprise, though, it is a complete failure. The cost of transporting a person or a piece of space equipment first to French Guiana and then into outer space on a nanocarbon cable ends up being even higher than the cost of using a normal rocket. The French public, who financed the whole project with their taxes, are also aggrieved that such a marvelous French invention is not even available to gaze upon in their own country. The American public has their Kennedy Space Center. The Russian people have their Vostochny Cosmodrome. But French citizens have to traverse the Atlantic Ocean to South

Paris 2121

¤ ¤ ¤

America before they can view the space adventures going on at their national spaceport.

So, in an effort to increase the commercial viability of the project and appease the French taxpayers, the space elevator is moved. Over the course of many months, the base station at the bottom end of the cable is loaded onto a massive boat and floated ever so slowly to La Havre on the French coast, then up the River Seine to Paris. When it finally gets there, the French will at last be able to rejoice in its grandeur (even though very few of them will be able to pay for a ticket to ride on it).

However, late in 2121 something goes dramatically wrong. Due to either a design fault or some accident, the entire space elevator starts to vibrate and shudder, before falling spectacularly down to Earth over Paris.

So the question is this: What's this got to do with ecotopia? Well, right before the collapse of the space elevator, a sturdy security robot was scheduled to load a plutonium battery into the elevator so that it could be transported to serve an orbiting spaceship. Luckily, the robot developed a glitch seconds before completing its task and automatically shut down. The elevator rose without the battery, and because of this, the entire population of Paris was saved from a much larger catastrophe: being rained on by a huge cloud of radioactive fallout.

Perth 2121 ¤ Life in the Post-Carbon Age

Some cities in the developed world are manicured to within an inch of life. They have precisely cultivated trees lining quiet sidewalks fed by impeccably timed water sprinklers. Perth, the capital of Western Australia, is one such city. It is a suburban utopia of green lawns and clean streets with fresh new cycleways and a light railway to boot. There are plenty of highways and cars and traffic lights in Perth, too, but do not worry: all are well-monitored by a citywide computer system to keep the cars moving around in an energy-efficient and time-saving manner.

The Perth city government also invests resources in educating Perth residents about how to curate their own lawns in an eco-friendly manner, including information regarding the best time of day to water the lawn so that the water isn't just evaporated by the glaring sunshine and what type of grass is more resilient for residential use. Perth then publicizes to the world how Green it is.

Perth's well-kept lawns and suburban roads lie in stark contrast to its long-term future. The Perth climate is getting hotter and drier and the soils thinner and more saline. Heat waves arrive so often that they become the new normal. To fight

against climate change—and make huge amounts of money at the same time—the Western Australian government will probably soon approve the opening of uranium mines all across the state. The uranium will then be used, they will argue, to power "climate-friendly" nuclear energy stations around the globe.

The scenario depicted here is not so optimistic about Western Australia's uranium future. It suggests that sometime in the late twenty-first century, one particular uranium mine hundreds of miles away from the city ends up having its huge tailings dam breached, thereby contaminating the rivers that flow through Perth with vast amounts of radioactive sludge. Many people in the city are forced to evacuate or choose to leave for other towns rather than face the high risk of health problems.

Alas, despite this uranium push, global climate change is not halted, and the city of Perth is not only contaminated with long-lived radiation but also ravaged by climate change disasters, including extreme drought, devastating wildfires, and disastrous soil loss. When water comes, it arrives in flash floods, which not only sweep away homes and roads but end up washing more contaminated sludge into the city. Any human survivors will have to downscale or de-develop the city, and they will have to learn to live on remaining local resources in order to survive.

In Perth 2121, two reliable resources are Balga grasstrees and mud. The remaining residents build their houses out of mud, sourced from the uncontaminated pockets of nearby earth and mixed with rocks, shells, fish bones, and dried plants. For strength, as well as decorative appeal, the dwellings are biomorphic, being inspired by the skeletons of local sea urchins. Alive, sea urchins are covered with an array of formidable spines. But when desiccated and denuded, the striking radial patterns of the sea urchins are exposed. With warm, contaminated seas lapping at the future Perth coast, there may be plenty of sea urchin skeletons lying around to serve as inspiration for the sturdy structure of the mud huts. However, there are no smart water sprinklers and no lawns, but the front and backyards of the huts are much more diverse and interesting to look at and play within than what preceded them.

The sea urchin mud huts might serve as infrastructure, but the economy in general is based on the cultivation of Balga, a type of grasstree adapted perfectly to dry Australian environs. The Balga can supply the practiced cultivator with all manner of products. The resin, for instance, can be used as an adhesive in toolmaking, the floral nectar provides a sweet drink, and the floral spike can serve as an effective fishing spear.

As for transport, the Post-Oil Age in Western Australia will allow the reemergence of a once popular form of Australian desert transport, the camel. Camels were imported from Arabia to Australia in the nineteenth century for transport and heavy work in the rural areas. But when trucks and trains came along and the camels were no longer needed, several thousand were released into the wild. Currently, there are three-quarters of a million camels eking out a living in the Australian outback. By 2121 AD, after Peak Oil, it will be time to enlist camel power once again for both eco-friendly transport and heavy lifting. Adorned with solar panels to power all manner of simple electronics as well as cooling and heating devices, camel transport will be organic and sustainable but also open to appropriate small-scale technology.

Nowadays, camels are employed in desert regions around the world. In modern-day Africa and Asia, they transport materials and medicines and they are used for security and policing operations, as well as for food distribution. In Perth 2121, caravans and squads of camels will take on these roles and more, acting as mobile libraries, roving schools, ranging health clinics, moving markets, and shifting security. Many will also serve as low-cost taxis, transporting passengers of lesser means to friends and facilities around the noncontaminated areas of Western Australia.

Ecotopia 2121

¤ ¤ ¤

Philadelphia 2121 ¤ "Plants Are People, Too!"

On a sunny July morning in 2121, a group of high-spirited citizens is arguing with city officials in the park at Philadelphia's Independence Hall. The subject of the argument is a tree, a chestnut painted with a big red circle, occupying a corner of the park. The circle signifies that it is scheduled for immediate felling.

The tree's ongoing existence, in a place where the likes of George Washington and Benjamin Franklin regularly strolled, has been argued over in the press and in official meetings for a year or two. Some officials declare that the chestnut is infected with some grotesque plant disease that might spread to all the other trees. Other officials report that the chestnut has been dropping its leafy limbs on the ground far too regularly, endangering passersby. Many Philadelphians did not buy into these stories of the risks posed by the chestnut and thought the stories were made up by a property developer trying to open up the area for a new project.

The chestnut had been living on its corner since the founding fathers were drafting the Declaration of Independence in what is now known as Independence Hall more than three hundred years ago. One eccentric old Philadelphian was so outraged by the impending execution of the chest-nut that he walked to the park at five in the morning to chain himself to the tree. The city workers didn't arrive until 9:00 a.m. and the city officials not till 10:30 a.m., so there was plenty of time for reporters to set up their equipment in order to capture the proceedings. The old guy, standing bound to the tree, is making an impassioned speech to the whole city:

This tree's even older than me. Why do we protect three-hundred-year-old buildings and not three-hundred-year-old living beings? We need trees to survive. Who do you think gives us the air that we breathe? So here today, on this auspicious site, I call for a Declaration of Interdependence. We have no right to kill this tree. It is part of our community. Plants are people, too!

By midday, the tree's felling was officially postponed pending an investigation. By late afternoon, the old man was being interviewed by statewide media. By sunset, thousands had come to honor the chestnut in the park and celebrate its survival. By late evening, engaged citizens had planted trees all around the city at sites suspected to be subject to dubious development plans.

During its colonial days, Cambodia's capital city, Phnom Penh, was known as "The Pearl of the East" because the French architecture on the Mekong riverside merged in beauteous contrast with the ancient feel of the traditional buildings. Today, Phnom Penh is as far from utopia as one might imagine. The city suffers from health-hazardous air pollution and intractable water issues, as well as urban squalor, overcrowding, misgovernance, corruption, and a general lack of social and environmental well-being.

The motorcycle traffic alone, considered by some to be iconic and full of character, is dangerous, smoky, noisy, and chaotic. There are no public buses on the city's roads, no trains to serve the suburbs—no public transport at all. The only visible state-sponsored infrastructure projects are new parliamentary buildings for the political elite, a few massive modern-style condos for the business elite, and a huge casino resort for rich Chinese gamblers.

One of the persistent challenges for cities on the Mekong River, including Phnom Penh, is flooding. These floods can be seasonal or episodic. The seasonal floods are generally welcomed by farmers but are still challenging for urban centers, while the episodic floods are often quite disastrous. With foreign aid, the Cambodian government has attempted to implement flood-control structures, but these are incomplete and far from reliable. When there are floods that sweep through the city and suburban streets, the government usually denies it is to blame and instead points a finger at the citizens because they have settled in the wrong places and because they recklessly dispose of their plastic bags, which clog up the waterways and impair smooth drainage.

Although it is now a city of two million, Phnom Penh was once purposely depopulated. During the rule of the Khmer Rouge in the 1970s, the urban middle classes were forcibly ruralized to work on the land to provide food for the nation. The ideology espoused by the Khmer Rouge insisted upon the primacy of agricultural production, both as an economic sector and as an ethic, and they believed that many people in the city needed to be weaned off their counterrevolutionary impulses by toiling with common folk in the fields. When the Khmer Rouge lost power in 1979, the city of Phnom Penh bounced back to be repopulated by displaced residents.

In Phnom Penh 2121, farms move from their

rural setting and come to the city (rather than the other way around), as the era of urban agriculture dawns. The importance of urban agriculture for developing nations like Cambodia is multifarious. By incorporating large-scale urban-based agriculture that works within the prevailing ecosystem, food security is enhanced and the nutrition of the urban poor can be improved. With food so close to its consumers, the costs of its production and transport are lessened. It's also likely that less food will be lost through spillage and decay, because the supply chain will have been shortened and simplified. It's doubtful that urban agriculture in Cambodia or elsewhere will ever completely replace rural agriculture, but it will complement it and so allow for a general increase in the efficiency of the national food system as a whole. The urban economy will also be boosted since, apart from the farming, many allied activities and services will be opened up—for instance, animal health services, bookkeeping services, and low-tech and high-tech transportation services.

The architectural background to this new "city of agriculture" involves the use of condo-style residences placed on a set of twenty-five concrete stilts above the flood-prone areas of the Mekong River. The river is not engineered and controlled but allowed to flow in a natural way through and around the city. Between the sinuous channels and water zones, the city residents have wet gardens to grow crops. During seasonal and episodic floods, Phnom Penh residents stay dry and safe above the flood line. The main crop species of the city is likely to be flood-tolerant rice, but many other wet plant species and freshwater crustaceans can be farmed as well.

Another advantage of this flood tolerance is the positive impact it has on biodiversity. In the early twenty-first century, Cambodian wetlands and mangrove communities are suffering the consequences of pollution and rapid development. Concrete channels, barriers, and dams often do nothing but make this worse, especially for the aquatic fauna. Here in Phnom Penh 2121 AD, however, the natural shape of the land and the use of an organic, wetlands-based agriculture encourage the riverine fauna ecosystem to flourish so that sustainable harvesting of the fish is made possible as well.

In the Cambodia of today, the public's socioeconomic and political expectations are rising. Along with other Southeast Asian citizens, the people

Phnom Penh 2121

□ □ □

of Phnom Penh aspire to better living conditions, especially with regard to housing. In many Asian nations, the condominium tower block has become the preferred standard of living for many city dwellers. In Southeast Asia, the word *condominium* usually refers to a high-quality tower block rather than a specific ownership scheme. The convenience and comfort of condominiums are reinforced by the provision of in-house shopping and some common space within which residents can relax and congregate. Such high-density residential tower blocks are also declared to be a lot more environmentally friendly than a typical detached or semidetached house.

After one or two "eco-condos" are constructed along the lines of the technical features outlined here, and after they are shown to survive and prosper in spite of the frequent episodes of flooding, they will serve as inspiration to construct even more such housing projects in Phnom Penh, until they become the ubiquitous form of residential living.

Ecotopia 2121

◻ ◻ ◻

Pittsburgh 2121 ¤ The Blissful Commute

Pittsburgh in the twenty-second century has become a network of pedestrian walkways. The walkways allow citizens to traverse their surroundings easily and comfortably on foot or with a wheelchair or perambulator. With creative bylaws, each walkway allows for a pleasant trip to work in a community atmosphere, where commuters can take time out for breakfast or dinner at walkway take-outs and rooftop cafés. Many Pittsburghers believe that a ninety-minute walk to and from work—stopping off to chat with friends and colleagues while listening to sidewalk musicians—is a much more preferable way to commute than being stuck in a traffic jam for an hour and a half.

So how does such a vast cityscape of interconnected pathways possibly come about? Slowly and gradually. First of all, the new bus lanes (along with improved quality of service) encourage more people to opt for public transportation—so much so that the public bus business becomes both fashionable and profitable. City planners also find ways to link bus routes with new city center walkways.

Over time, it becomes evident that every yard of new walkway reduces motorized traffic proportionally, since people decide to leave their cars at home and get about on foot. This may not be a conscious decision on their part but an unconscious acknowledgment that the city is somehow more accessible and enjoyable when the traffic jams and parking problems can be avoided.

Over many years, public support for walkways encourages more investment. Also, since the health of the people is improved as a result of all the physical exercise they are now getting, the government is less burdened with a variety of health care costs, and this frees up the funds for investment in additional walkways. Eventually, by 2121, because of all the mini-cafés, arty food bars, and colorful community kindergartens, the walkway environment spreads back down to colonize the streets as well.

Plymouth 2121 ⌘ City of the Neo-Luddites

It is a cool, still night in early November 2111, and our location is the port city of Plymouth, England. Arms, limbs, and heads are being blasted across the harbor by homemade explosives. The citizens in the city nearby are yelling joyfully.

The body parts belong to robots, and the people are hooting as the machines' dismembered remains get blown up and tossed out to sea. After watching the fiberglass robot heads bob up and down in the waves for a little while, the citizens all retire to local pubs to celebrate, proudly singing the first verse of a nineteenth-century poem penned by Lord Byron—to contemporary music, of course:

As the Liberty lads o'er the sea
Brought their freedom, and cheaply with blood,
So we, boys, we
Will die fighting, or live free,
And down with all kings but King Ludd!

Exactly three hundred years before this night of revel and singing, on a chilly November evening in 1811, near the textile factory towns of Nottinghamshire and Yorkshire, the Luddite movement was born. A group of textile artisans, skilled weavers who were about to be laid off, went into a factory at night and smashed all the machines.

Today, in the early twenty-first century, anybody who resists technology in any way is often branded with the Luddite label, which is meant pejoratively. Bill Gates's lawyer, for example, used the term in this way in the courtroom against the people who filed lawsuits claiming Microsoft was breaking monopoly laws.

Yet the modern-day myth that Luddites are backward techno-haters just exposes a willful ignorance of history on the part of those who use it as a term of abuse. The original Luddites were not opposed to *all* technology. They were against the uncontrolled and unnecessary use of it to let businesses make minimal savings on labor costs. As they explained in the *Nottingham Review* in 1811:

If the workmen dislike certain machines, it was because of the use to which they were being put, not because they were machines or because they were new.

It's also the case that the quality of the product, the cloth that was spun by the new machines, was demonstrably inferior to the cloth fashioned by the human artisans—thus disenfranchising consumers as well as the workers.

As they made their feelings known and put their ideas into action, the Luddites and those who supported them in any way were actually very much risking their lives, since the Parliament in London had hastily passed a law imposing the death sentence on anybody found guilty of Luddism. About seventy Luddites were hanged, and many others were sent to detention in penal colonies abroad. To be a Luddite was to be deeply courageous.

In addition, the government sent not police or other law enforcers to deal with the situation but instead dispatched twelve thousand soldiers. Ludicrously, these soldiers were marched around the textile counties of northern England in a high-profile attempt to deter Luddite activity. Given that this was 1811, at the peak of the Napoleonic wars in Europe, it is surprising that so many soldiers could be found in England and then wasted on such an expensive task, but such was the hostility toward workers' rights among those in power. The Luddites lived during a time when workingmen and -women did not have a vote or a political party to represent them, and when unions were outlawed. All this means that the smashing up of a machine seemed a perfectly rational action in order to keep one's job. The historian Eric Hobsbawn calls Luddism "collective bargaining, early nineteenth-century style."

Various public show trials were held in towns and villages in England to warn people off of supporting the Luddites. If Bill Gates's lawyer had been around in the nineteenth century, maybe he would have been on the side of the parliamentarians, allied with rich merchants to force workers out of jobs and calling for them to be hanged if they resisted.

◻ ◻ ◻

Not all parliamentarians in London were out to vilify the Luddites. This is why we find the joyous people of Plymouth in 2111 singing aloud the poetry of Lord Byron. In 1811, Byron wrote and published his first major collection of poetry. It made him into a huge celebrity around the nation, but as a life peer in the House of Lords he was also obliged to turn up in the upper house of Parliament once in a while to vote on matters of state. One of his first speeches in this venue was an eloquent statement against the death penalty for Luddites. From this speech he later crafted his "Song for Luddites," the second verse of which reads:

Plymouth 2121

◻ ◻ ◻

When the web that we weave is complete,
And the shuttle exchanged for the sword,
We will fling the winding sheet
O'er the despot at our feet,
And dye it deep in the gore he has pour'd.

And why would people in Plymouth in 2111 so happily chuck robot parts into the sea? Well, the robots had just arrived in a huge container ship, all preset and preprogrammed to take over their jobs. The people foresaw that the robots would be employed to load and unload the ships and would move goods from here to there all over the city. Soon enough, robots might end up teaching the kids at school and pouring the tea and beer in the cafés and pubs. This prospect of total automation in Plymouth had galvanized Plymothians into action, hence all the flying robot limbs.

Of course, a robot company isn't going to get very far in the twenty-second century just by advertising the superiority of robots over humans. So in Plymouth the robot company spent lots of money trying to seduce and manipulate the populace into loving robots by making the machines as human as possible, giving free pet robots to kids, and donating robots to the local schools.

These freebies soon ended up floating on the sea as well.

Because the owners of the robot company couldn't manipulate Plymouth as they had wished, they soon left the city, and the Plymothians had to resurrect the operations of the port themselves using old-fashioned labor and machines. Over the course of the next decade, they also realized how much better off they would be if they rid their workplaces not only of robots but of many other unnecessary technologies. They would be much safer and lead much more satisfying lives with less need for capital investment, which in turn made them less dependent on external influences. Many people also felt the city to be more peaceful and family friendly. Welcome to Plymouth 2121.

Like Nottinghamshire in 1811, Plymouth in 2121 is not a wholesale reaction against every form of new technology but resistance to those forms that end up being used to make local people weaker and relinquish control over their work lives. Occasionally, the Plymothians are warned that they will never compete with other English port cities on the global stage, yet Plymouth 2121 manages to secure sole trading rights to the eco-cities of Athens and Antalya and also to the Slow City of

Malaga, exporting and importing all they need to survive and prosper.

¤ ¤ ¤

If Luddism develops in such a way as to become the most popular form of politics and management in Plymouth 2121, how can it be classed as a type of ecotopia? Plymouth's Luddism involves the enshrinement in law of the Precautionary Principle, which states that if there is any risk to human or environmental health from a new machine system, and if this risk is uncertain or makes for possibly irreversible consequences, then Plymouth will err on the side of caution and not just willy-nilly adopt the machines. All the risks of any new machines are then weighed and counterweighed by a democratically elected jury, including those recruited to specifically speak for the marine and terrestrial wildlife in or nearby the city. Only after substantial evidence is found that no harm will befall Plymouth, either its people or its environment, will new machine systems be approved.

Most Plymouth citizens accept this as standard, since they feel there's no need to rush a technology into use just to make some anxious investor or corporation happy. Plymouth will survive well enough without certain machines for as long as it takes to ensure that they are safe. If any developer attempts to circumvent this rule by invoking a British law not voted for by a local MP or by appealing to an out-of-town magistrate, then the bylaw states that any registered Plymouth residents may undertake eco-sabotage in order to ensure the health of the city, as long as they publicly announce their actions by singing aloud all three verses of Byron's "Song for Luddites." Here's the final verse:

> *Though black as his heart its hue,*
> *Since his veins are corrupted to mud,*
> *Yet this is the dew*
> *Which the tree shall renew*
> *Of Liberty, planted by Ludd!*

Plymouth 2121

¤ ¤ ¤

Prague 2121 ✕ "There Are No Roads in Bohemia"

In the nineteenth century, the struggling artists of Europe's big cities often grouped together in low-rent neighborhoods to live a lifestyle described as "Bohemian," derived from the supposed origin of Gypsies in the province of Bohemia. The Bohemian lifestyle rejected conventionality and celebrated personal freedom, usually through some form of art. It also involved vagrancy and voluntary simplicity and a willingness to undermine the supremacy of an obnoxious concept called *the work ethic.*

Prague is the capital city of the province of Bohemia, within the modern nation of the Czech Republic. In the scenario presented here, Bohemians—of the lifestyle kind—have traveled from around the world to be in Bohemia, the geographical locale. They are lured there by the self-described desire of Prague to be a center for world art. Where Paris and London might attract artists with ambition or entrepreneurial talent, Prague 2121 attracts artists with spirit, with a sense of humor, and with an intellect. When the Bohemians of Prague do manage to sell a piece of work, the funds are used to pay their back rent, and anything left over is spent with their friends over the course of a single night. Since they reject consumerism and volunteer to live simply, they do not use up superfluous resources nor produce much waste. As the Bohemian lifestyle becomes fashionable, more of Prague's citizens see that life can be more satisfying with fewer material possessions.

There are no roads in Bohemia. So said Gellet Burgess, a famous lifestyle Bohemian based in California during the early twentieth century. Many people take this phrase figuratively to indicate that each person has to find his or her own road in life. But here, in Prague 2121, it is interpreted literally. The roads of Prague, once filled with sad, lonely people in noisy private cars, have been disrupted and subdued by landscape artists who have closed off the roads, one by one, in overnight landscaping frenzies, pedestrianizing and "socializing" the tarmac into new urban parks and plazas. So modified, they become more suitable for artistic pursuits—and for parties.

Prague 2121

▫ ▫ ▫

Puno 2121 ¤ City of the Lake, City of the Cloud

On the shores of Lake Titicaca, thirteen thousand feet above sea level, an enchanting city of one hundred thousand people is nestled. The city is noted for textile crafts made from alpaca wool and for a magnificent condor monument that overlooks the lake. Welcome to Puno, the folkloric capital of Peru. Puno is way up in the Andes, thousands of feet higher than the old Inca capitals of Cuzco and Machu Picchu. Here, the air is thin, there are few trees, and there are also few cars—partly because the area between Lake Titicaca and the mountains is so minimal there is hardly any space to build a road.

Today, Puno is known as "The City of the Lake." Offshore, on Titicaca's surface, there are floating artificial islands made of reeds. These islands are handmade by the Uros people, and on them they build their village homes, which are also handmade from the reeds. The Uros originally adopted this style of architecture for defensive purposes, since in centuries past they were able to evade capture by Inca armies when they floated their villages out in the middle of the lake.

Because they have lived on the lake like this for centuries, the folklore of the Uros is imbued with stories about floating, escape, and freedom and survival. Puno 2121, as presented here, is an extension of these ideas. A floating settlement emerges above the clouds, fashioned from the textiles woven out of alpaca wool. Although the Uros believe Lake Titicaca to be sacred, they also acknowledge that it is rapidly deteriorating. The water level is receding every year because of decreasing annual rainfall. It's also becoming polluted, as cities around the lake dump more wastewater into it every year, degrading both the reeds and the fish stocks. To lessen their impact on their sacred lake—or to escape the attention of new armies—some of Puno's Uros residents, as imagined in this scenario, rise into the skies in a new settlement that floats high in the clouds, in the realm of the condor.

Puno 2121

◻ ◻ ◻

Rekohu is a small, isolated South Pacific island six hundred miles off the eastern coast of New Zealand. It has a population of about one thousand people of three ethnic backgrounds: British, Maori, and Moriori. Rekohu is a place unique on Earth. It is possibly the last island in the world to be discovered and colonized by humans. It is also a place of unique biodiversity, with a host of fish and bird species that can be found only in this one place. One of the unique species is the Chatham albatross, an elegant seabird with a beautiful dusky crown and an orange flare on each cheek.

The sea provides much of the economic sustenance for Rekohu's human inhabitants, but because of overfishing and inappropriate net laying, many native fishes in and around the island's lagoons have become rare and endangered. Another problem is that due to coastal erosion (both natural erosion and manmade) the structure of Rekohu's largest lagoon is bound to change over the course of the twenty-first century, becoming more and more silted up. This may very well contribute to radically endangering the biodiversity of the lagoon, perhaps pushing some species toward extinction.

The Rekohu islanders are very sensitive to these extinction possibilities, so in this scenario for the future, the island's largest lagoon is converted to sustainable aquaculture. Here, within a network of causeways, new sea farms are organized, where fish and shellfish are cultivated in an eco-friendly manner. Some segments of the network also help preserve wild fish by allowing their young to shelter and grow before they reenter the wider areas of the lagoon or the open sea. With luck, the network of causeways will also stabilize the processes of erosion and siltation, and so help fish species escape extinction.

At the center of the network is a newly formed lagoon city. This is Rekohu Te Whanga 2121, population between two and three thousand. This new city has agencies devoted to three particular sectors: a peace center, a wildlife center, and a maritime industries center. All three have facilities devoted to teaching various skills to locals and visitors from islands around the Pacific Ocean.

In the past and in the present (and in the future, too), Rekohu is irrevocably tied to the ideals of pacifism. All of the residents in Rekohu Te Whanga 2121 take a pacifist pledge, and they customarily use the Moriori greeting *me rongo*, meaning "with peace." The Chatham albatross has been adopted

Rekohu Te Whanga 2121

☒ ☒ ☒

Reno 2121 ⌘ Geothermal Capital of the West

Reno, Nevada, has a well-known history regarding two social matters, gambling and splitting. The gambling revolves around slot machines, dice, and playing cards, while the splitting revolves around the legal act of divorce between the partners in a marriage. In both these activities, Reno was way ahead of its time compared to the rest of the United States. By the 1930s, it had become the gambling and divorce capital of the country.

In Reno 2121, the themes of gambling and splitting are combined in a program to develop an inexhaustible energy source located nearby: geothermal energy. Geothermal energy is a renewable energy produced by transferring heat from hot subterranean rocks up through geological splits in the ground to steam-driven power plants on the surface. If discovered and efficiently tapped, the heat from these geological splits could provide perhaps a quarter of the American West with all the heat and energy it might need forever into the future, with not so much as a whiff of hydrocarbon emitted or radiation leaked.

However, geothermal prospecting is a somewhat blind endeavor. There is a lot of risk and uncertainty regarding where exactly geothermal vents are hidden. In this respect, geothermal energy is even more uncertain than oil or gas, which at least may exhibit surface features, such as oozing crude, that might point toward an opportune prospecting area.

In twenty-second-century Reno, high-rolling gamblers are enticed to play a new game: sink the geothermal drill. Each shot at the game costs about $10 million and involves picking a spot somewhere in the outskirts of Reno and sinking an exploratory drill. One of the reasons for the high price of a bet is that geothermal activity is associated with granitic rock, and drilling through granite is a very tough and drawn-out affair. However, high rollers who strike it lucky will be given a stake in an endless resource and can also be happy in the knowledge that they helped Reno 2121 become the Geothermal Capital of the West.

Reno 2121

□ □ □

Resistencia 2121 ¤ Wild Feminism

Resistencia, in northern Argentina, sits on land once inhabited by the native Guaykuru Indians. Their resistance to European settlement is honored in the name of the city, which was eventually founded in the late nineteenth century by Italian immigrants. Resistencia is classified now as "subtropical," but with global warming, by 2121 it has become decidedly tropical.

Resistencia has chronic water problems—sometimes there's not enough, sometimes there's too much—and always it is contaminated. Activists blame this on a litany of *isms*—capitalism, industrialism, cronyism, for example—as well as anthropocentrism: the belief that humans are the center of the universe and other creatures don't matter so much. In Resistencia 2121, another *ism* is highlighted: namely, sexism. Women in Resistencia are stuck with less money, education, and political influence than men. Every day, they bemoan the dirty water gushing from their taps, and every day they protest the diversion of river water from their garden crops into the city's messy factories. For decades, the women of Resistencia have campaigned to have clean water piped to every home and garden, but each time they've been thwarted by chauvinism, corruption, and the threat of violence.

Now, in April 2121, after months of drought, when water flows only to the homes of the rich, they finally mobilize. Luckily, they have a band of water science students in their midst, and the deft skill of these so-called watermaidens—swashbuckling throughout the city's infrastructure—is used to bring city leaders to their knees. The watermaidens divert water supplies from the rich suburbs and into the houses of the poor. At first, the city leaders call in a militia to deal with the situation, and a band of gun-wielding men gather on one bank of the Rio Negro determined to storm the occupied water center on the other bank.

The autumn air is hot, as usual. The autumn water is warm and slow flowing—but this is not usual. The men stride into the river forthrightly, but halfway across they are attacked by fish biting at their legs and ripping at any exposed flesh. They have no idea at first what is happening, for these waters are usually too cold for piranha, but now, as indignant screams of agony echo across the banks, the fishy surprise attack forces the men to retreat. Listening to the cheers of the watermaidens on the other side of the river, the city leaders recognize that they have no choice but to negotiate.

Resistencia 2121

◻ ◻ ◻

Rio de Janeiro 2121 ⌘ The Sea Towers of America

When the land now settled as Rio de Janeiro was explored in 1506 by mapmaker Amerigo Vespucci, he characterized the Indian societies living there in utopian terms. While wary of the strange behaviors of some of the natives, Vespucci wrote with admiration of their stateless liberty, communitarian spirit, and freedom from overlords. These descriptions are said to have had an influence on Thomas More's original *Utopia*, published a decade later.

Six hundred years after the time of Amerigo Vespucci, Rio de Janeiro 2121 is making a new attempt at utopia by constructing enormous sea towers to shelter eco-neighborhoods, which are "future-proofed" against rising sea levels and coastal erosion. The citizens of each sea tower can choose to operate either as a self-dependent economy or as a trading economy. In either case, each community is able to independently draw up their own guiding values. Some sea towers will likely choose values such as equality, self-sufficiency, or religious commitment, while others will espouse a balanced "work-life complex," or cultural diversity, or sexual liberty. It's assumed that people will prefer to live in towers with others who share their values, but no sea tower government will be granted permission to expel those who do not strictly adhere to these values. If they try to do so, there is still recourse to Brazilian law.

The initial capital cost will be covered by the federal government, but after the sea towers are set up and occupied for a few years, the financial cost of maintaining the infrastructure will be covered by the community itself. No private money will be allowed to be invested or loaned to the community, which will permit the community members to live without a debt burden and undue corporate influence. There will, however, be a means test to ensure that all aspiring residents are from a low-income background; those who exceed a certain (community-agreed) limit must forfeit their excess income to the community or move back to the mainland. This will incentivize the tower residents to live in a sharing mode, discouraging poverty and encouraging a community atmosphere. It will also ensure that each sea tower lives within its ecological limits.

Rio de Janeiro 2121

☐ ☐ ☐

In the thirteenth century, when Francis of Assisi was a mere youngster at a ball among the daughters of noblemen, he was asked, "Now, Francis, will you not soon make your choice from these beauties?" He replied, "I have made a far more beautiful choice. Whom? La poverta!" Though his family was wealthy, Francis chose to eat and sleep with the poor on the streets of Rome and Assisi.

Besides choosing a life of poverty, Francis was renowned for his kindness to animals. He would speak to them in a personable way, in the same manner he'd speak to people. It has been said that Francis also enjoyed preaching to flocks of birds and packs of wolves in the wild. Moreover, the birds and the wolves would listen attentively.

Maybe these stories are apocryphal, but it doesn't really matter, for many Catholic faithful believe them to be true. Every October 4, the day after Francis died, Catholics from around the world gather to honor him by bringing their animal companions to church for blessing. In 1979, Pope John Paul II pronounced Francis the patron saint of ecology.

Most religions have to be dragged, screaming and kicking, into ecological consciousness. The same can be said of Catholicism. However, the new pope is a keen follower of Saint Francis; indeed, he adopted the saint's name when he was elected in 2013. Within days he was declaring that the pursuit of riches leads to despoliation of nature and that it is the sacred duty of Catholics to preserve all living parts of God's creation. Many reactionary Catholics threw their heads back in disgust.

If the impetus of Pope Francis's Green message flows throughout wider Christendom, the ramifications may be profound. In this scenario, we may eventually arrive in Rome 2121. Here, Catholic priests from around the world have come to emulate various Green Catholic heroes, and every Sunday after Mass, priests join with schoolchildren to get their feet dirty by walking the city picking up trash and planting trees. Another development: in 2121, the papal head becomes a permanent representative in the United Nations General Assembly. If he's anything like Francis I, the pope of 2121 may become the single most powerful environmental lobbyist in world politics.

Rome 2121

¤ ¤ ¤

Salto del Guairá 2121 ¤ City of the Waterfall

The Guairá Falls of the Paraná River in Paraguay were one of the natural wonders of the Americas. Their seven columns discharged the highest volume of water of any falls in the world. Their cacophonous roar could be heard twenty miles away. For many years it was a tourist attraction—a special favorite among the locals—for its power, glory, and the bellowing mists sent into the pristine landscape.

That was until 1982, when the military government blew away the rocks over which the water fell to create a reservoir for the newly constructed Itaipú Dam. The local indigenes, the Guarani, and the Paraguayan mestizos as well, prayed and mourned as their sacred falls died.

Salto del Guairá, the city, still exists in Paraguay despite the demise of its namesake landmark and despite the loss of the one million tourists who passed through it each year before the dam was built.

In the future, both the town and the falls have reemerged in splendiferous style. Perhaps the Paraguayan military bombs the Itaipú Dam by mistake, or perhaps a series of natural disasters strike it and make it tumble—a great earthquake, for instance, or a series of storms and floods. Or maybe the dam just crumbles from old age or neglect. For whatever reason, someday in the late twenty-first century the Guairá Falls are reborn in a glorious cascade. By that time, the Guarani have gained control of their land and the Paraguayans have mustered democratic resistance to their army's constant interference in government. Together, the Paraguayan people and the native Guarani set about rehabilitating the falls with a scenic city. And once again, millions of tourists are very happy to come and see the falls.

Salto del Guairá 2121

◻ ◻ ◻

San Diego 2121 ⌘ Deep Blutopia

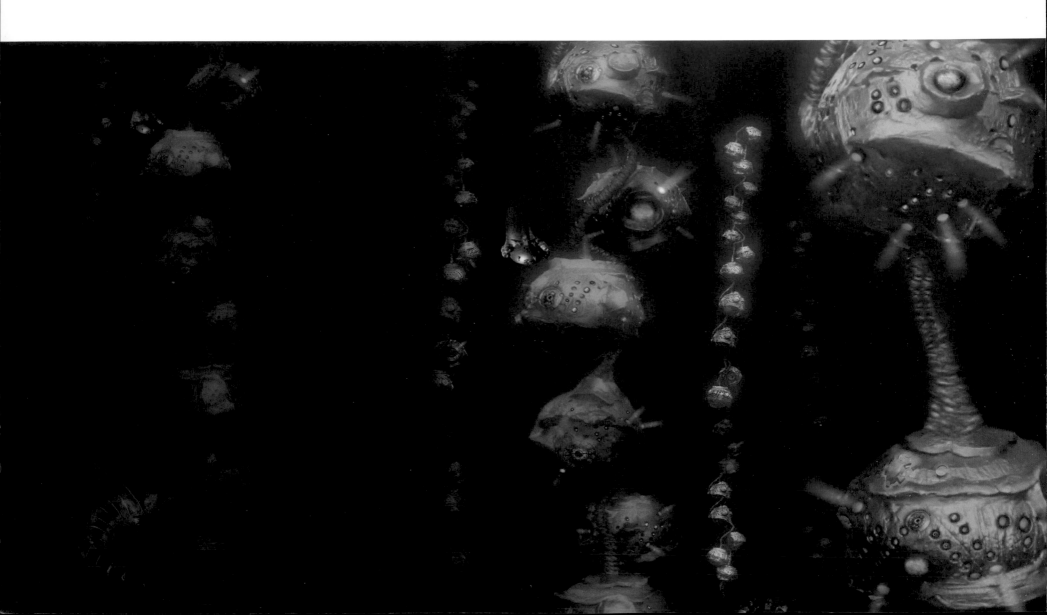

Within a decade or two from now, an "ocean grab" seems to be in the cards as nations like Russia, the United States, China, Japan, and Canada position themselves to exploit future resource possibilities in the world's seas. Some envision that the oceans will become the new frontier and be colonized in the same way as most other parts of the Earth in earlier centuries.

Amid this growing interest in oceanic development, and in response to global sea-level rise, some cities may reassess their relationship with the oceans. Such is the case in San Diego 2121. By early next century, the rising sea level may drown or erode much of San Diego. This would make living and working there so troublesome and risky that many families and businesses might well venture to move to a more stable environment just offshore under the sea.

In this design, a suburb of San Diego is resurrected in 2121 within the sea. Using the technology developed and perfected as a result of one hundred years of oceanic exploration, as well as from the naval experience of San Diego's submarine fleet, these submarine homes—suitable for about ten people each—are tethered, like giant California kelp, to the rocky seafloor using robust holdfasts.

The idea of such a sea-bound community is not just to hedge bets so that San Diego could survive a drastic sea-level rise, but also to promote eco-friendly lifestyles in the sea using traditional and novel techniques of mariculture. It will nevertheless be an expensive enterprise, but there's a commercial impetus from the growth of the seaweed biofuel industry, which by the early twenty-second century might fully replace fossil fuels.

San Diego 2121

¤ ¤ ¤

San Francisco 2121 ¤ The City of Growhemia

If Prague 2121 is the capital of *Bohemia*, then San Francisco 2121 is the capital of *Growhemia*. Here, neighborhoods transform themselves into proud "self-grow" eco-villages with intense social interaction based not on the exchange of goods and services but primarily on the exchange of ideas and experiences. Growhemia is heavily into urban gardening, and its citizens work tirelessly to share the latest ideas about growing plants for food and for fun.

Like their artistic brothers and sisters in Bohemia, the Growhemians are untroubled by social disapproval as they set up Green markets in the streets to trade necessities without paper money being involved. Some days, organic carrots are the benchmark currency; other days, it's Californian beets. Genetically modified foods are also traded, but their value is very low because they must be transported in expensive, heavily armored crates to prevent interaction with organic crops. Most Growhemians have abandoned cars decades ago in favor of pedestrianism. While the journey to Growhemia might seem like a lifestyle choice, it is driven just as much by practicalities. The food sold in many twenty-second-century supermarkets seems to cater only to rich carnivores and those with private vehicles (which are also prohibitively expensive to run).

Where Bohemians live for their fascination with art, Growhemians live for their fascination with gardening. Like Bohemians, they have to be thrifty and rely on social networks, not exchange networks, to get projects finished. However, everybody quickly realizes that this is an immense advantage, since working with other people who share the same passion provides for abundant pleasure—and no one has to pay for entertainment, because daily life is always entertaining. The Growhemians are very much attuned to their own place and their own time. Most of them realize the planet is in trouble, but they have no pretense that they can save the entire world. All they can do is survive within, then beautify and savor, their own small part of the world.

San Francisco 2121

▢ ▢ ▢

San Gimignano 2121 ¤ The City of Organic Saffron

San Gimignano in Italy is a unique place. The town's ten thousand citizens are clustered onto a Tuscan hilltop setting that has changed little in six centuries. Among the old stonewall houses and cobblestone streets are a dozen medieval towers, all built during the town's heyday in the thirteenth century. After about 1400 AD, the nearby city of Florence exploded onto the European trading scene as a major financial center. Because of this, San Gimignano was very abruptly overlooked and subsequently ignored as wealthier townsfolk and tradespeople moved to Florence. The fact that San Gimignano was suddenly "left alone" like this means that, to modern-day visitors, a trip to the city is like a trip back to the Middle Ages.

Despite missing out on modernization, San Gimignano has a thriving economy revolving around old-time crafts and traditional agriculture. Most notable among these are saffron cultivation, wool craft, and Vernaccia wine making. The production secrets for these goods are rather ferociously guarded to ensure they cannot be copied. As you might imagine, San Gimignano is not the sort of place where innovation is as much respected as tradition.

One innovation that the residents are particularly hostile toward is GMOs (genetically modified organisms). By 2121, just about every habitable place in the world has been infiltrated and invaded by one GMO or another that has gone feral. GMO animals have escaped from labs into cities, GMO plants have escaped from nurseries into the wilderness, and GMO crops have escaped from farms to drift far across the countryside. In 2121, only a few places in Europe remain certified GMO-free, and San Gimignano is one of them.

Gene-tech companies don't like to use the word *mutant*, but biologically that is what GMOs are. Here in San Gimignano 2121, strict biosecurity measures are in place so the mutants can hope to see the town's pretty architecture only from another Tuscan hilltop, many miles away.

San Gimignano 2121

◻ ◻ ◻

Santiago 2121 ¤ City of Copper Blue

Santiago is a city of five million at the foot of the Chilean Andes. Just outside the city limits is a copper mine. The mine releases a slow trickle of sulfuric acid and arsenic into the outer neighborhoods of Santiago. The residents have kicked up a fuss for years, saying this toxic stuff is getting into their water, their soil, their air. The city government is sympathetic, but the national government says that the factory isn't even *in* Santiago, so residents there should stop complaining about it.

For years into the future, the two levels of government are likely to argue back and forth over the issue. The city government requests that the city limits be extended to encompass the mine so they can regulate it. The national government defends the status quo, saying the mine provides employment for Chileans from all around the nation and that the mine company pays its fair share of taxes. They also tell Santiago's environmentalists that they should thank the mine for stopping Santiago's urban sprawl. Because the copper mine is such an economic asset for Chile (a special minister was appointed to handle its affairs), the government and the mine seem tied in lockstep. So the copper mine stays operational, all the while killing nearby forests, poisoning wildlife, polluting streams, and sending a foul stench over suburbs.

One day in 2101, the copper mine finally runs out of copper. It's only at this time—when all the promises about rehabilitation have to be fulfilled—that the national government finally rezones the mine site to be within Santiago city, telling the city that it's now their responsibility. Santiago residents are unfazed. They are just thankful the mine has shut down, and over the course of the next twenty years they work to fill in the tailings dam, clean up the toxic residues, and replant the Andes forests with healthy native trees to make it a place worth living. By 2121, a suburban village has been erected for the descendants of those once adversely affected by the mine, and a new railway connects it to the center of the city. The whole thing is judged a success when the first *pudú*, a beloved native animal usually too timid to descend from the pristine forests of the Andes, is sighted in the village, seemingly at home in the outer suburbs of Santiago 2121.

Santiago 2121

⊡ ⊡ ⊡

São Paulo 2121 ¤ The Salad Revolt

The crowded, noisy streets of Brazil's largest city, São Paulo, are far from utopian at present. For eight million commuters, each working day involves many hours of torturous travel on overheated, overpriced, and overcrowded buses. Private automobiles move no faster; they just clog up the streets. A few years ago, when the government announced with fanfare that it would be spending billions of dollars on stadiums, first for the World Cup and then for the Olympics, it was just too much for many Paulistanos to bear, and they revolted in the streets. Because many protesters were arrested for carrying vinegar (which can serve as an antidote to tear gas) with them, the movement was dubbed "The Salad Revolt."

As an intermediate response to the Salad Revolt, the Brazilian government announced that billions of dollars—a figure on par with the cost of hosting the World Cup—would be invested in public transportation in the coming years. Perhaps, a few years down the road, private cars will be banned from entering inner São Paulo as well. Looking further ahead, all that space previously occupied by cars might be given over to urban agriculture. Maybe at first these gardens would be small-scale. Families would grow vegetables in nearby city streets and on highway islands, so they wouldn't have to travel three hours to work at a job just to get money to buy food.

At some point in the late twenty-first century, perhaps during a particularly poor season for the Brazilian national soccer team, great runs of tarmac will be dug up by guerilla gardeners, who claim that food is more important than football. Soon after, the highways and byways are planted with verdant rows of diverse food crops that stretch as far as the eye can see. The Salad Revolt will then be complete.

São Paulo 2121

□ □ □

Shanghai 2121 ¤ Gay City of the East

For many decades, China's central government attempted to control the population of its cities through a one-child policy. The city authorities believed that urban overpopulation presented them with large-scale problems associated with

- land and water scarcity
- overcrowding
- air and water pollution
- waste management issues
- food supply problems
- health and education deficits
- transportation and energy supply problems

Many Chinese planners consider it sustainable if the population growth rate of an overcrowded city is kept at zero via stringent family planning. Over time, though, a host of social problems have emerged:

- *Human rights abuses:* Some believe it is unfair that any government should stipulate how many children a couple can have, and they bemoan the overbearing and invasive nature of provincial "baby police," who knock on doors with monotonous regularity to count family members, track menstrual cycles, set fines, and impose sterilizations and abortions.

- *Gender bias:* The chauvinism and sexism in Chinese culture have promoted a preference for male children. This means both abortion and infanticide have often been rampant in Chinese cities, as couples act to make sure their firstborn is a boy. This gender bias has also led to an ongoing skewing of the sex ratio, so that over the whole of China, men outnumber women by some twenty million.

- *Sociopsychological issues:* Having a city filled with only children has convinced some psychologists that cities like Shanghai are full of selfish, spoiled "little emperors" unable to adjust to communal settings with their peers.

- *Dwindling worker numbers:* There is the prospect that in the future the dwindling number of people of working age will negatively affect the economic status of China. This problem has prompted a change in government policy, and now China is introducing a two-child limit. This two-child policy may alleviate some problems of the one-child policy, but it will nevertheless keep family planning as the province of the authorities rather than the individuals concerned, and anyone wishing to have children will need a permit for each of them.

Despite the issues listed here, some environmental scholars outside of China also advocate some form of population control. The future scenario painted here as Shanghai 2121 takes into account this population pressure but proposes three other measures—a mixture of the social, the technological, and town planning—to push Shanghai toward ecotopia:

1. The outlawing of heterosexuality
2. The compact city design ideal
3. The construction of a network of piezoelectric pedestrian tubeways

Outlawing heterosexuality: Throughout the time of communist rule, China has had a troubled relationship with homosexuality—decriminalizing it only in the late 1990s and delisting it from a compendium of psychological diseases only in the early 2000s. In ancient China, attitudes toward homosexuality were much more accommodating and perhaps quite tolerant, and many emperors were held to have practiced it at some time or other. Such tolerance might well be worth resurrecting for the benefit of Shanghai 2121. In order to lower the birthrate in Shanghai to a point where environmental pollution and resource supply become manageable, homosexual marriage will be encouraged and heterosexual marriages (as well as heterosexual physical relations) are to be prohibited. Given that the men outnumber the women, this might be a blessing for the "leftover" men. And given the sometimes strained and unfair relations between Chinese men and women, this may be a blessing for the women as well.

Many scholars sympathetic to the advancement of gay rights have tried to put forward the idea of the *natural* or *unchangeable* nature of a person's sexuality in order for their sexual status, be it homosexual or heterosexual or some mix of the two, to be acknowledged, respected, and accepted in modern society. But many scholars provide evidence that sexuality is mutable and that, with the appropriate encouragement—social, erotic, and legal—Shanghai citizens can be encouraged to accept and embrace homosexuality as right and proper, both for themselves and for the social and physical environment of their city.

Compact city design: This urban design concept promotes high-density working and living arrangements within a layout that encourages both easy and efficient public transportation and the promotion of pedestrianism and urban cycling. In a *compact city*, every resident lives close enough to public and private amenities—schools, shops, clinics, entertainment centers, government offices, and so forth—so they need never think of using a private vehicle. This means also that the city of Shanghai in 2121 will be a low-cost, low-energy-consumption, and low-pollution type of city. And it will mean a very high degree of social interaction (resulting in more business opportunities for its citizens and also, probably, more security against crime, as there are always people around to observe and maintain safety). There will also be less urban sprawl, so that the surrounding countryside will remain intact and green spaces in the city will be preserved.

Piezoelectric tubeways: The piezoelectric pedestrian tubeways depicted for Shanghai 2121 will provide all-weather pedestrian transportation citywide. The surface of each tubeway will harness the pressure of step energy, converting it into productive electricity. Some of this electricity will be used to service the tubeways' energy needs (for lighting, water pumping, waste disposal, etc.), but any surfeit will be credited to the energy account of the person doing the walking. Thus, you can pay for your own electricity bills just by walking around Shanghai, and the more you walk, the greater the credit you can earn. Indeed, professional walkers might earn a livable wage if they are allowed to swap their credits on a free market. A beneficial side effect will be to improve the health of the Shanghai populace, saving the government lots of money while improving the quality of life.

Shanghai 2121

◻ ◻ ◻

Sharjah 2121 ¤ The Statue of Arabia

"Jullanar. I see Jullanar. I see the sea-girl!" shouts a young boy excitedly as he leans over the side of a small dhow bobbing gently upon the sea. He's there, just off the coast of the Emirati city of Sharjah, with his grandfather, a fisherman—though long since retired. They've quietly escaped together from the family home in a rented boat for an afternoon of secret fishing. However, as is usual for this part of the sea, there are no fish.

If Abu Dhabi's "little brother" is Dubai, then Dubai's "little brother" is Sharjah. The cities are packed tightly right next to each other on the Emirates coast, and Sharjah likes to thinks that whatever Dubai can do, so can Sharjah. So in the mid-twenty-first century, when the rulers of Sharjah see the rulers of Dubai building impressive, great towers and palaces on engineered artificial islands, the Sharjah rulers think: *We can do that. We will do that!*

Alas, Sharjah's magnificent marine towers are just as prone to failure—through both structural degradation and financial woe—as those of Dubai. So in the Sharjah of the first years of the twenty-second century, there exists a huge, decrepit monument decaying off the city's coast. It was planned as the massive Statue of Arabia in the sea, the Emirati version of New York's Statue of Liberty, which would let Sharjah residents and its visitors bask in a sense of splendor as they looked out from their balconies facing seaward. Instead, it is a crumbling pile of concrete ruins destroyed by bad planning and bad weather.

But all is not lost. There is ecological salvation from this degradation. For many years, the cities on the Persian Gulf have been destroying coral reefs and marine communities with busy ports and oil spills as well as with all the engineered islands. But here the abandoned Sharjah statue is to become a place where coral can form, oysters can mature, seagrass can develop, fish can breed, and a patch of mangrove can grow.

It's around this place, one day in early 2121, that we find the young boy and his grandfather in their dhow. They were in a state of serious relaxation, just watching the clouds drift along above, breathing in the balmy air, waving at passing boats. They gave up on the fishing, since there was nothing alive in the water. Or so it might seem. Suddenly, the boy cried out excitedly, pointing at an undulating shadow in the water: "Jullanar. Look! Grandfather, I see Jullanar!" The fisherman was a bit startled before he saw what the boy was pointing

Sharjah 2121

¤ ¤ ¤

to. Then a huge smile crossed his face. Jullanar is the name of a mermaid from an ancient Arabian myth. The grandfather knew there probably wasn't a mermaid in the water—but the reality was almost as amazing: a dugong, a shy, peaceful marine mammal very similar to a manatee, the likes of which the grandfather had not seen in the Gulf waters since he himself was a boy.

For grandfather and grandson, the sighting was truly remarkable, but the two of them together were hardly believed when they went home that evening and told their story to the rest of the family. By the twenty-second century, dugongs are generally acknowledged to be extinct in the Persian Gulf, and the boy's father and mother, and his brothers and sisters too, all just laughed at them. However, grandfather and grandson were convinced of what they had seen, and the two of them vowed to sail again out near the statue in the sea to try to sight the dugong once more. This time they'd aim to get a good photograph of it to show the whole family.

¤ ¤ ¤

Later that evening, perched on top of their roof overlooking the darkening sea, the grandfather recounted to his grandson a half-forgotten story that had been passed down from ancient Arabia. It told the tale of Jullanar and Abdullah:

Abdullah is a fisherman in olden times, and he and his wife are blessed with a bevy of children in their small, modest home. As he sets out to fish one morning, his wife tells him that another child is on the way. Unfortunately, Abdullah is not so blessed when it comes to catching fish, and for months his luck has been against him and he comes home with a paltry catch or nothing at all.

Today is another bad day. Abdullah has not caught a single thing. Once again, like many times before, he stops off to see his friend, the baker, to sit and chat for a while and then ask for any spare bread so he can feed his children that night. The next days are the same. Hours are spent on the sea. No fish are caught. He visits the baker on the way home, chats with him quite a while over tea, and then asks for help. The baker is always kind in response.

Once in a while, during these tough times, Abdullah can be found on the rocks by the sea weeping silently and praying, wishing desperately for some few fish to come to his net. His despair is such that one day he falls dejected into the sea to drown his tears. After a short moment of panic, Abdullah is calmed by a touch on his shoulder. He spins around to see a woman floating below the surface of the sea with him. She is breathing water as though it is air, and the bubbles wash over him and into his lungs. Abdullah breathes the bubbles in, and miraculously, he finds he too can breathe the water. The woman is Jullanar, the mermaid, and she swims below into the deep and gestures for Abdullah to follow.

In the wake of the mermaid's flapping fluke, Abdullah follows her to a grand city, deep underwater. The mermaid takes him around the city, and everything seems strange, as though it's an inverted copy of what is common on land. Here, people are happy to care for fish, not to catch them, and the sheer joy of caring is sustenance enough to live on. Here, also, people work in cooperation, with no fighting, and no war and no leaders and no kings; and nobody has any money and nobody wears any clothes.

Another inversion that suits Abdullah is that the underwater city is festooned with myriad precious jewels, all sparkling in a hundred shades of blue and rose. They are strewn all around the city, willy-nilly and completely unsecured, as though they have no value beyond their shining beauty. Jullanar notices Abdullah's fascination with the jewels and, at the end of her guided tour around the city, she gives him a few radiant pieces. Abdullah thanks Jullanar with a smile and swims to the shore.

Abdullah climbs onto land then rushes with the jewels to his friend, the baker. The baker welcomes him with tea and sits down for a chat. Abdullah offers the baker the small set of jewels. "Oh, thanks, Abdullah. What are they, my friend?"

"They are jewels from the sea. I give them to you as payment for all the bread you've given me."

"Thanks, Abdullah, but the chats with you over tea are payment enough. You are my friend, and I always have bread to spare. And besides, my friend, these 'jewels'—they just look like pebbles."

Abdullah looked at the jewels again and saw the baker's words seemed true. They were pebbles. Quite pretty pebbles, to be sure, but just pebbles. Abdullah was puzzled, for they looked so much

Sharjah 2121

□ □ □

Sinaia 2121 ¤ Recycling in Transylvania

Amid the post-totalitarian landscape in Romania are remnants of half-finished, grandiose megaprojects. These range in kind from crumbling power plants and decaying factories to silted-up shipping canals and weather-beaten Communist Party monuments. The scenario here is a pointer to the possible use of the physical remains of such projects, a compact vertical suburb in the Transylvanian city of Sinaia made from the remains of an abandoned hydropower facility, rising from the misty mountain valleys into the air. This is no planned community of towers but more of an unplanned, organically organized, communal dwelling fashioned in an informal way over many years, then refined, rearranged, and elegantly sculpted. The towers are a testament to the power of recycling.

In the eastern half of Europe there are some eight million Roma people, an ethnic group often referred to as Gypsies by English speakers. Romanians are usually at pains to make sure foreigners understand that Roma and Romanians are ethnically distinct. The former have origins in ancient India, and the latter have a mixed Roman and Thracian heritage. According to popular prejudice in Eastern Europe, the Roma are responsible for inordinate damage and destruction to public buildings, including their own state-funded houses. Another popular story about the Roma is that they are dealers of garbage. More often than not, this story is cast about to disparage the Roma lifestyle, which is often segregated from the mainstream wage system. In Sinaia 2121, a time and place where environmental sensitivity is much higher than today, garbage recycling has become a more honorable activity than it is currently. The success of the recycled towers suggests that the Romanian government should hand over the decaying infrastructure of its abandoned white elephants to the nearby Roma people. This will provide an opportunity for the Roma to show that they can build communities in their own way. They can surely do no worse than Romania's government.

Sinaia 2121

▫ ▫ ▫

Singapore 2121 ¤ Sky-High above Asia

The Lion City was born to trade, and the little problem of planet-wide sea-level rise shall not stop this. In Singapore 2121, the city floats above the eroded Singapore Island on a disc-shaped "space station"–type platform. Astronautics experts the world over have claimed for years that large space stations, the size of small cities, can be constructed both in space and on planets in such a way that they function as independent ecosystems, recycling their own air and water and producing their own food. A few "practice space stations" have been built on Earth to do the same thing, such as the Eden Project in England and the Biosphere II project in Arizona. Fans of these projects usually claim that research into such artificial ecosystems has practical application to Earthly situations. From such projects, and similar ones developing in Singapore, the Singapore Space Agency may be well equipped in the future to engage in the construction of Singapore 2121 as depicted here, especially given Singapore's advancing interest in other ostentatious space projects, such as the Singapore Spaceport, which aims to give Singaporeans adventurous rides into space at the cost of a few million dollars per ticket.

So what has to change before Singapore can get from today to Singapore 2121? The answer: almost nothing. Megaprojects such as these will suit the showy authoritarian style of Singapore's government, for it is a virtual one-party state, where the People's Action Party leads and others must follow. Singapore just has to keep getting wealthier and invest in ever more elaborate technological megaprojects year after year in order to

1. Train its engineers in the requisite skills to develop airborne cities
2. Showcase advanced technologies as the true manifestation of progress and civilization
3. Make Singapore stand out from its Southeast Asian neighbors (both literally and figuratively)

These things, done in concert, will keep the Singaporean citizenry in political lethargy believing that their nation is on the right track to a glorious future.

Currently, a new wave of techno-megaprojects is sweeping through Singapore, for example:

1. Singapore plans to incinerate all its waste and dump the resulting ash into the sea so that hundreds of acres of land are available for new "eco-towns." (Yes, there will

probably be coastal pollution and habitat destruction.)

2. Singapore plans to build the largest seaport in the world. (Yet only a tiny percentage of the exports and imports will have Singapore as their origin or final destination—the port is just a stopover hub for ships from around the world.)

3. Singapore plans to build hundreds of huge caverns three hundred feet underground to store all the undesirables of the city— things like risky chemicals and explosives, and some garbage—to keep the surface area more "Green and livable." (Yes, the construction process will involve billions of tons of global warming gases being produced and risk contamination of the groundwater.)

It is likely that most of Singapore's currently planned megaprojects, especially the ones portrayed as Green, will not stop global warming or sea-level rise and may actually contribute to making these even worse. In fact, as the decades of the twenty-first century come and go, the world's changed climate may make the future environment quite horrific for many people living in Southeast Asia, with floods, droughts, water shortages, and energy shortages becoming more common. The wealthiest Singaporeans will expect to be able to buy their way out of the entire environmental crisis. But, as the twenty-first century dawns, the poorer workers, the unemployed, and the large immigrant population will either have to lower their lifestyle expectations quite drastically or move elsewhere. In the worst-case scenario, the impending global sea-level rise will erode their island home into the sea.

Singapore today is a city of four million people, but the floating bubble city of Singapore 2121 will be able to serve as home for only twenty thousand. Therefore, this utopian Singapore of the future will be able to house only the nation's business and government elite. Over time, everybody else—if they don't have the money or contacts— will be forced to emigrate.

So why would such a place be considered a utopia? Currently, Singapore's elite have to spend a lot of time and effort to systematically keep in check the activities of workers' unions, immigrant nongovernmental organizations, civil rights groups, and opposition parties, a process involving the passing of biased laws, ever-present police activity, defamation lawsuits against critics, and crackdowns on antigovernment speech. Besides this, the elite of Singapore find the common workers' complaints about the lack of social welfare and public health care in Singapore very annoying, as the 2006 Wee Shu Min scandal made evident. In this incident, the daughter of a parliament member berated an individual from the "sadder classes" before telling him to "get out of my uncaring elite face." A number of People's Action Party politicians issued halfhearted apologies, but many others insisted that she was just telling it like it was and anybody from the underclasses who was offended just could not face the truth. In Singapore 2121, there will be no more pesky working-class or middle-class people to be annoyed by or who will constantly need to be kept in their place. The elitist utopia of Singapore 2121 will replace the lower and middle classes with robots and computers. The robots won't demand high wages or health care—or political reform—so the business/government elite will live in a luxurious paradise. Not only that, but everyone on the Malayan peninsula will look up enviously at Singapore floating so high, and this will put even more smug smiles on the faces of Singaporeans.

Singapore 2121

◻ ◻ ◻

Sochi 2121 ⌘ The Capital of Circassia

Sochi is the largest city on the Caucasian Riviera, abutting the eastern side of the Black Sea. It's the warmest city in the entire Russian Federation, a place where Russians go for summer holidays and family fun at the beach. For most Russians, Sochi was a rather peculiar place to be selected for the 2014 Winter Olympics, since they know it as a warm seaside town. In the winter, though, it can get icy and snowy, and the Caucasus Mountains fold gently around the town.

Today, in the early twenty-first century, Sochi residents are having their front and back gardens slip away from beneath them. The Russian authorities, at the bidding of President Putin, were in such an unyielding hurry to make certain the city was ready for the 2014 Games that they built the new roads in haste throughout the city and surrounding countryside and in the mountains nearby. The new stadiums, hotels, sports facilities, and ski resorts were also quickly built. The local residences were often perilously close to the construction. Now, when it rains, the land slips sometimes and forested hillsides have smashed down onto a number of neighborhoods. If they're lucky, a family might find only a fallen tree in their backyard. If they're unlucky, their whole house will have slipped down the side of a hill or a mountain.

Landslides are not the only environmental problem. By the time the Olympians had packed their skis and headed home, Sochi had been afflicted with the following:

- *The loss of many wetlands* buried under both the Olympic Village and the rubble left over from the village's construction.
- *The deforestation of eight thousand acres of the Sochi National Park* for the construction of skiing and shooting facilities.
- *The tainting of the water supply* due to the illegal dumping of rubble in a water-protected zone. Some neighborhoods were told by unidentified officials not to drink the water. When they asked why not, they were told that it contained dangerous substances.
- *The initiation of avalanches* induced when ski slopes were laid out and forever altered the mountain environment.
- *The destruction of pristine natural settings.* In order to build ski facilities, ice fields in the mountains were destroyed, the forests denuded, and swamps filled in. The whole hydrological system of Sochi has changed,

Sochi 2121

¤ ¤ ¤

causing unforeseen flood events as well as local desiccation. Those parts that were wet have become dry, and those parts that were dry have become sodden all year round.

- *Wildlife disturbance.* Many kinds of animals, from bugs to birds to bears, have fled to unknown locations in search of new homes. The sites of some of the Olympic venues were once a stopover zone for migrating birds, but now they are devoid of them.
- *Chronic corruption.* The construction sector and local politicians have been implicated in corrupt activities on numerous occasions, and local residents have been illegally evicted from certain prime real estate areas.

In addition to all these problems, one particularly destructive road was cut through pristine woodland to connect the city with a new, luxurious private villa built for President Vladimir Putin high in the Caucasus Mountains. Worse yet, the Russian leadership has made it clear that anyone who speaks out against any of these problems will feel its wrath. Putin had a faithful "Sochi posse,"

which aggressively pursued environmental activists to prevent any demonstrations that might have interrupted the Games, including jailing the most vocal of them for indefinite periods for very minor infractions such as "swearing in public" or "being a nuisance."

As if all that were not bad enough, the 2014 Olympics served as a grand diversionary tactic for Putin as he amassed Russian troops to invade Ukraine just a few hundred miles away on the north side of the Black Sea. The moment the Games ended, he ordered them into Crimea.

¤ ¤ ¤

Those residents of Sochi who survived the Olympic environmental onslaught just about caught their breath, but now construction has resumed for both the FIFA World Cup and the annual Formula One racing series. More vegetation is to be removed, more coastline despoiled, more environmental activists secured behind bars. This is the present and the near future of Sochi. Meanwhile, the people of other Black Sea countries such as Georgia and Moldova live in fear that Russia will use the World Cup to divert international attention from Putin's preparations to invade their countries.

Luckily for Sochi's beleaguered citizens—and for Georgia and Moldova, too—an unlikely environmental hero is emerging in the form of a bright blue-and-yellow parrot named Max. Max is really quite an average pet parrot, friendly and talkative, but unlike most parrots he has a luxury apartment in Trump Towers, Manhattan, all to himself. (Well, all to himself in addition to a couple of feline roommates, but they get on pretty well by all accounts.) Fortunately for Max, he doesn't have to fork out the $17,000 per month rental fee for his expensive roost, since it's paid for by the ill-gotten gains of a guy named Chuck Blazer.

Blazer is Max's owner. He is—or was—also the US Secretary of FIFA, and he's very wealthy. According to the FBI, Blazer gained his riches through a great, messy bunch of corrupt deals he made with FIFA and the sports minister of the Caribbean nation of Trinidad.

Whether it was the parrot that chattily let out the secrets of the corrupt deals is not known, but eventually the FBI found out about the decades-long misdealings. It seems that every four years, one or another of FIFA's representatives bribed other FIFA representatives in order to secure World Cup hosting rights. As the investigation proceeds, the concealed records of Max's keeper are bound to expose the purported bribes at the heart of the Russian 2018 World Cup hosting bid. Because of the FBI investigations, FIFA is under pressure to take the 2018 World Cup event away from Russia and let another country host it. If this happens, then the environmentally damaging construction in Sochi will grind to a halt immediately. On the other hand, Putin will be absolutely livid. Hosting the FIFA World Cup event is a prize so glittering that if he loses it, he'd take it very personally and might even send tanks into Georgia and Moldova out of spite, taking over every Black Sea port in a weird macho attempt to save face and prove his potency. Putin's aides might try to calm him down by promising they'd work to make him *president for life*, for instance, and telling him that the city of Sochi would be renamed *Putingrad* in his honor after he dies.

◻ ◻ ◻

When speaking of Sochi's future, it should be noted that there is a dark history to Sochi that

Sochi 2121

◻ ◻ ◻

few Russians know about. Sochi is not a Russian city. It became part of imperial Russia in the late nineteenth century through the genocidal murder and displacement of a native people known as the Circassians. About three million Circassians were living in and around Sochi in the early part of the century, and they regarded themselves as belonging to an independent state that comprised a collection of tribes, with Sochi as their capital. In their own language, the Circassians referred to themselves as *Attéghéi*. The name is believed to derive from *atté* ("height") to signify a mountaineer or a highlander, and *ghéi* ("sea"), together signifying "people dwelling in the mountains near the sea."

As Russian imperialism grew in the nineteenth century, Britain and France tried to help the Circassians to resist Russia by supplying arms and munitions. However, after the Crimean war of the 1850s, Western European willingness to further help Black Sea tribes was sapped of impetus. Under these circumstances, during the 1860s a crazed Russian military commander named General Yevdokimov rampaged up and down the coast near Sochi, slaughtering Circassians of all ages, burning their churches and mosques and shrines, destroying their towns and villages, and devastating their crops and livestock. By 1864, the general had taken Sochi. Today he is honored as a war hero by the Russians of Sochi.

Merely waging a war against the Circassians was not enough, however. The Russian minister of war, in 1860, drew up plans to eliminate the Circassians altogether, to either "kill them all or push them out into the sea." These plans, which were nothing short of ethnic cleansing, were published widely for all Russians to read and were endorsed by Tsar Alexander II. Many Circassians chose to resist, and they died for it—some four hundred thousand of them. Other Circassians believed their only chance of survival was to submit immediately to forced deportation. With Russian soldiers pointing guns at their backs, almost the entire Circassian population—millions of people—were moved on horseback or by foot to the ports of the eastern Black Sea, including Sochi. There they were crowded onto ships and sent off to Turkey, hundreds of miles to the south. Many reports from the time suggest that nearly half of them died of disease and hunger or exhaustion on the way. The Turkish port masters noted how the ships arrived one by one like floating graveyards from Sochi.

Today, 150 years later, about four million ethnic Circassians are living in different nations of the extinct Ottoman Turkish Empire, mostly in Turkey itself but also in Jordan and Egypt. There are also a few thousand Circassians still in Russia. Many Circassians still identify strongly with Sochi as their homeland and feel aggrieved when it is touted as a great Russian city by the world's media. Circassians outside of Russia are trying to highlight the fact that the Olympic Games and World Cup stadiums are being built on sites where their people were massacred (but those Circassians within Russia are subject to arrest if they attempt the same).

As the decades of the twenty-first century come and go, the geopolitics of the Black Sea is bound to change, maybe gradually or maybe via a series of crises. One day, though, Russia may well be depleted of easy oil and gas, and any new military forays in the Black Sea aimed at strengthening its international standing may do just the reverse—costing it both diplomatically and financially. It's quite possible that worldwide sympathy, and trade, will flow to the multiple smaller countries around the Black Sea rather than to Russia.

That noted, we can look forward to the far future of Sochi and describe a scenario of Circassian revival there. If the smaller nations around the Black Sea, such as Ukraine, Georgia, and Turkey, do become wealthier, and if they work cooperatively in business and diplomacy with the rest of Europe, then a Circassian uprising against Russian domination is possible. Here we arrive at Sochi 2121.

When the Circassians do make their way back to their home city, they are likely to find an industrialized, polluted city named Putingrad. However, this will hardly deter the Circassians from rejoicing at being home again. It will take a generation probably, but they will slowly reinvent Sochi as their capital.

One part of this project might be the resurrection of traditional Circassian theology—which mixes monotheism with ancestor worship. Central in this theology is a providential god, *Tha*, who expects nothing in return for the universe of goods he provides. Although Tha is not a judgmental god, the spirits of dead ancestors are thought to be, since they hold sway over everything that a

Sochi 2121

¤ ¤ ¤

Circassian person does in his or her lifetime. The Circassian ancestors from the nineteenth century will probably look upon the Circassians of Sochi 2121 with grace and happiness, pleased that their descendants have recovered their homeland. But, also, the ancestors will want the Circassians of 2121 to do good things with their renewed capital, to respect Sochi as a place to forever call home, and to make an honorable living there—that shows due regard for the mountains and the sea.

Ecotopia 2121

◻ ◻ ◻

Springfield 2121 ¤ On the Road to Nowhere

In the early twenty-second century, the Springfield city administration is running out of money. The highways, for instance, are in a state of disrepair and haven't been resurfaced for years.

When the residents congregate for public events, they often speak their mind with regard to the highways: *Let them degrade. We don't need them. We're not in any hurry to leave. It's so nice here!* And, compounding this infrastructure problem, they also declare that they don't care if it takes visitors an extra-long time to get to Springfield, because that means only the ones who really like the place make the effort to come.

The desire to balance the books by cutting expenditures on the highways has the side effect of making Springfield even nicer. There's less pollution, less noise, and less roadkill. It might mean that the prices of fuel, food, and beer imported into the town will rise, but then again Springfield has a positive take on these issues, too, declaring that with fewer highways and Springfield being so nice to walk around nowadays, they'd be inclined to drive less and wouldn't need to use so much gas. As for the food and drink, Springfield 2121 can easily find space within the city—now that all the highways have disintegrated—to grow their own healthier food and maybe even start an organic brewery.

Another thing that makes Springfield 2121 so much nicer is that the nuclear plant has closed down. It was scheduled for refurbishment, but when the residents learned that their electricity bills would go up by a whopping 100 percent to pay for this, they just packed up all the uranium rods and sent them off to Siberia, where they'll get dumped secretly and cheaply somewhere near Orenburg.

Stuttgart 2121 ¤ The Art of the Grapescape

Ecotopia 2121

¤ ¤ ¤

Stuttgart is where the automobile was invented at the close of the nineteenth century. One hundred years later, the city still hosts the headquarters of Mercedes-Benz and Porsche. But one hundred years into the future, given the decline of fossil fuels and the diminished value of cars, and in concert with cities like Springfield and São Paulo, great ribbons of tarmac in Stuttgart's suburbs will be dug up by guerrilla gardeners, who claim that good food is more important than excessive mobility. By 2121, vineyards and vegetable plots will re-color the city from shades of gray into shades of green. Some people will object, of course, to this total Greening of suburbia, but many officials will see value in supporting these activities, especially if it means they get more votes. One particular horticultural output, the grape, used both for wine making and for food, takes over from the car as the iconic Stuttgart product. Within a generation, Stuttgart becomes famous for its urban grapescapes. And by converting a portion of the now defunct roadways into vineyards, intermixed with horticultural plots, the following eco-benefits will accrue:

- The grapescape will decrease *food miles* (the distance a food product must travel through its line of production).
- The grapescape will lower the ecological footprint of each food product (that is, the total environmental impact of each food).
- The grapescape adds greenery to the city, providing for a nicer environment and reducing harmful water runoff and soil loss.
- The grapescape can contribute to social inclusion and health (when the neighborhood comes together in a commercial or collective way to produce healthy salads and an income for the community).
- The grapescape provides for the productive reuse of urban wastes (both solid and liquid) as they are fed back onto the landscape to enrich the soil as fertilizer.

The city of Stuttgart does not ban cars altogether. It bans only mass-produced cars from factories. Those who want cars must build them from scratch by themselves from either sustainable products or salvageable parts. This practice liberates the locals from being mere consumers in a global market dominated by multibillionaire corporations and makes them producers of their own technology, so that they are much more involved in the creative processes.

Sydney is regularly voted one of the "most livable cities" in the world. However, the environmental problems of this livable city are rife, and they are getting worse with every passing year. Wildfires, droughts, flash floods, dust storms, sandstorms, hailstorms, tree loss, insect plagues, bad air, bad water, bad soil. In the near future, it's likely that terrible one-hundred-year floods will occur once every few years, and the beautiful Sydney harbor will be eroded away, silted up, and contaminated to a level where neither humans nor fish can swim in it. Rising sea levels will wipe away Sydney's ports and the famous sandy beaches as well, while overdevelopment and urban sprawl—all the way to the Blue Mountains some thirty miles away—further degrade the terrestrial and riverine environment.

Long ago, before industrialization, the skies of Sydney were astonishing in their blueness. This was not only because of the fresh ocean vista and cloudlessness but also because the eucalyptus trees in the Blue Mountains released naturally scented terpinoids into the air that cast a shimmering blue haze. Today, the Sydney haze is brown and not of a pleasant scent.

The solution: blue environmentalism. The moniker "blue-green" is usually associated with eco-capitalism—namely, the belief that a highly liberalized market economy can work well to protect the world's ecology. In Sydney 2121, though, blue environmentalism denotes the total protection of the city's two most important environmental resources: the blue sea of the harbor and the Blue Mountains that mark Sydney's outermost boundary on land. If these two resources are preserved in an authentic ecological state, then Sydney's future is assured, too.

In Sydney 2121, liberal capitalism flourishes as it has since the city's founding in the nineteenth century, but under strict rules of investment whereby every individual in business, from the average CEO to the real workers, attends annual environmental education classes before being granted a license to work. There are also zones categorized as off-limits to development. For instance, no more construction is approved near the beach or in the mountains, or on sites of importance to aboriginal culture and myth. The aborigines of Sydney, the Koori, managed to live for tens of thousands of years without destroying the land within the city boundaries, and that's something all Sydneysiders can learn from.

Taipei 2121 ¤ The Austronesian Homeland

The vast majority of Taiwan's twenty-five million people are of Chinese descent, having arrived in successive waves of colonization from the seventeenth century on. Although Taiwan has been effectively independent since 1945, the Communist Party of China wants it to be brought within Chinese domain. The only thing halting a full-scale invasion, it seems, is the presence in the Pacific Ocean of Taiwan's ally, the United States.

This stalemate is bound to continue far into the future, but Taiwan will likely not sit idly by. It will need to develop new strategies of self-protection. One such strategy is depicted here as Taipei 2121, where legal action is taken at the United Nations. In a striking announcement to the UN General Assembly, Taiwan's government declares that all land rights in Taiwan will revert to their original "pre-Chinese" colonization status. Under this new legal regime, the land of Taiwan doesn't belong to Taiwanese of Chinese descent and certainly not to China, but instead it is the patrimony of the original inhabitants of Taiwan, indigenous tribes such as the Atayal and Amis, who occupied the island for eight thousand years before the first Han Chinese ever set foot there. Of course, the Taiwan government doesn't just hand the land over to the indigenous people—that would be political suicide—but it does sign a treaty stating that rent in arrears and into the future must forever be paid to the native communities by the Taiwanese state. All in all, this tactic increases the taxes per head by a tiny percent, but the legal barriers to Chinese invasion are greatly bolstered.

This stance has multiple effects. First, it strengthens the international conviction that Taiwan must forever be defended from China. Second, the indigenes of Taiwan, neglected for so long, suddenly have power over their own future, and they set about advocating traditional kinds of land management for Taiwan's capital city, in effect fashioning it as an ecocultural utopia. The success props up Taipei as an exemplar of First Nations around the world, and since Taiwan is said by anthropologists to be the origin of many Asian and Pacific peoples in the Southern hemisphere, it is granted honorary status as the Austronesian capital of the world.

Thimphu 2121 ✖ Constructing Shangri-La

Thimphu is the capital city of Bhutan, a small Himalayan kingdom wedged between India and Tibet. In the late twentieth century, Bhutan tried to give up on gross domestic product, or GDP, as the primary measure of the nation's well-being and replace it with gross national happiness, or GNH. The four pillars of GNH are (1) sustainable development, (2) preservation of Bhutanese culture, (3) preservation of the natural environment, and (4) good governance. These pillars, the Bhutanese leaders say, are more directly related to human happiness than any measure of finances and economics—so all new projects in Bhutan must be judged against these four pillars.

Many GNH supporters feel that, by using GNH, Bhutan will one day—maybe soon or maybe by next century—reach full utopian status whereby everybody lives in a state of happiness with each other and with their environment. This utopian impulse is advertised and promoted by Bhutan's leaders at home and abroad. When doing so abroad, it's common for allusions to be drawn between Thimphu and the fictional Shangri-La. Shangri-La was a hidden Himalayan paradise dreamed up by the British novelist James Hilton in the 1930s. It was a place where everybody was

happy and the people lived forever. Since then, Shangri-La has become a part of pop culture, seen in various movies and TV shows as well as holiday resorts, festivals, and theme parks.

However, there are many problems with the real version of this fantastic place. In the 1990s, a full sixth of Bhutan's population was forcibly expelled from the country because they were "not Bhutanese enough." Those expelled, the Lhotshampas, were descendants of Nepalese immigrants who arrived in the nineteenth and twentieth centuries to farm the lowlands of southern Bhutan. For many decades, these new Bhutanese lived in peace with the old Bhutanese, who themselves had migrated to Bhutan from Tibet about a thousand years before. In the late 1980s, the Bhutanese government enacted a "one nation, one people" campaign, making it illegal for the Lhotshampas to speak their own language or wear clothes associated with their heritage. Those who stood up against these biased laws, and many who did not as well, were either harassed into leaving Bhutan or forcibly expelled by the Bhutan army. For these refugees, the international fame of Bhutan as the "Land of Happiness" is a bitter pill to swallow.

Even for the citizens of Bhutan—the ones who are happy to dress in traditional Bhutanese costume every day of the year—there are many problems that detract from their happiness. Thimphu, for example, though better off than the rest of Bhutan, has poor educational standards, dilapidated housing areas, inadequate health facilities, and a woeful social welfare system—along with an overbearing, ethnically uniform, elitist, and authoritarian type of government. To aspire to be happy under these circumstances, and to celebrate "happiness" over all material concerns, has been highlighted by some social scientists as a ploy to make Bhutanese people content with their very modest lot in life and pacify political resistance. Maybe, also, this focus on "happiness" signifies a resignation that Bhutan can never truly develop sufficiently to be economically or financially strong. Like hippies in 1960s America, Bhutan seems to have given up competing and decided to "drop out" from the international economic rat race.

In a place insulated from the seductions of consumer society, GNH could perhaps be fully embraced by unwary citizens. The government of Bhutan had banned TV and the Internet until 1999. Now, though, that isolation is breaking up as more and more people have access to international broadcasts and electronic media, and there seems to be a rapid change in the desires and expectations among Thimphu's citizenry. The reality now, according to some Westerners who've conducted research there, is that the vast majority of Bhutanese are only paying lip service to GNH and "happiness" as they vigorously pursue the goods, services, and lifestyles they've seen on TV and the web.

So where does Bhutan go now? In this depiction of the Bhutanese capital, ecotopian ideals are achieved by fulfilling six commitments by 2121 AD:

1. A commitment to the wholesale practice of GNH, not just giving it lip service. The current GNH framework, though, is problematic from many angles, and it needs adjusting. First, it must be made measurable in some way through the drawing up of democratically agreed-upon standards and criteria. Second, the "preservation of Bhutanese culture" aspect of GNH has the tendency to encourage and support ethnocentrism. This theme is best abandoned so that individuals of Bhutan from different ethnic groups can decide for themselves what they value rather than having it prescribed by

Thimphu 2121

◻ ◻ ◻

243

an ethnically homogeneous elitist government. Third, to make GNH stronger and more worthwhile, an ongoing democratic debate about the meaning and policies of GNH must be enshrined in the constitution of Bhutan.

2. A commitment to reserve a proportional number of parliamentary seats for each ethnic group, for young people, and for women.

3. A commitment to an annulment of the king of Bhutan as head of state, since this position is far too much a symbol and base for the ethnocentric policies forced upon all the diverse people of the nation.

4. A commitment to expanding the construction of a real version of the Shangri-La idyll through social evolution (since it serves as an eco-friendly and utopian way to raise needed revenue from tourists that can then be used to fund GNH projects). If the Shangri-La theme is so endearing and useful to the leaders of Bhutan, then they should be encouraged to go full throttle with it. Instead of using it as an advertising slogan that pretends to describe present-day Bhutan, use it as an ideal future possibility that Thimphu's citizens can work toward achieving.

5. A commitment to total forest cover. At present, 60 percent of Bhutan's land is forested, down from some 74 percent in 1980. For Thimphu 2121, however, the forests are encouraged to recolonize just about all of Bhutan, and they are also introduced into Thimphu, to become part of the capital's cityscape as well.

6. A commitment to invite the expelled Lhotshampas back to Bhutan and to compensate them for lost land, and then include them in democratic governance, the reforestation effort, and the construction of a multicultural Shangri-La.

The primary change in the Thimphu social fabric, then, is the abandonment of the promotion of any "authentic Bhutanese culture" concept. In its stead, there's an elevation of an open and evolving Bhutanese "forest nature" concept. From this, the citizens of Thimphu are allowed to construct their own identities within an ethnically diverse and malleable Shangri-la fantasy and to connect this identity in myriad ways to the forests that surround them.

Timbuktu 2121 ⌖ Geoengineering a Desert Utopia

Some eight hundred years ago, Timbuktu was one of the wealthiest cities in Africa, rooted in place in the Sahel grasslands, with grand markets, forested glades, engineered waterways, and famous schools of learning. In Western countries it seemed shrouded in mystery, but maybe that's because it was a long way from the centers of European culture and Christians were generally prohibited from entering the city. Despite its illustrious past, Timbuktu is now a city impoverished by climate change—water-poor, baked by a relentless sun—and a place where it is increasingly difficult to eke out an existence.

Some governments around the world—Russia, China, and India, for example—are starting to believe that we don't have to give in to climate change but should develop even more powerful technologies to deal with it, something they call geoengineering. So, in the late twenty-first century, they spray millions of dense plumes of white dust into the global atmosphere, across vast swaths of barren land, and onto huge zones of polar ice, believing that this will cause excess heat energy to be reflected back into space. By the end of the century, nobody really knows whether or not this has worked. The climate disasters of drought and megastorms continue to manifest in Russia, China, and India, though the leaders there assert that climate change would be even more severe if they hadn't sprayed dust over everything. However, all this mucking about with the global weather ends up being a cataclysmic disaster for sub-Saharan Africa, since the air there becomes noticeably drier and the rainfall completely disappears, impoverishing the region's people even further. Once upon a time, the Sahel grasslands receded southward, away from the city, at a rate of only one-quarter mile per year, but after all the geoengineering, the grasslands recede at a rate of a dozen miles per year.

By way of an apology, Russia, China, and India decide to set up even more bizarre geoengineering experiments, basing their labs and facilities in Timbuktu. The people of Timbuktu have absolutely no doubt that their promises are completely useless, but they're willing to play along because the projects will bring money and employment to the city. This investment of resources also resurrects Timbuktu as a place of learning when a brand-new university is founded, which Timbuktu residents can attend free of charge. As the engineers undertake ridiculously expensive and futile megaprojects to induce rain clouds in the Sahara or bury atmospheric carbon dioxide in its rocks, the locals quietly continue to grow their crops as best they can in small-scale plots, hoping that, one day in the future, at least their children will be able to procure well-paying jobs in some other city.

Timbuktu 2121

¤ ¤ ¤

The Fukushima nuclear meltdowns in Japan in 2011 resulted in the abandonment of a host of small cities on the nation's Pacific seaboard. Tokyo, too, was under threat for a time, but luckily the radioactive plumes blew in another direction and Tokyo's twenty million residents were spared. In all likelihood, though, there will be two or three or more Fukushima-type events in Japan before we get to the close of the century, and in this scenario, almost the entire population of Tokyo has to be evacuated because yet another "unpredictable" seismic event hits yet another "perfectly safe" nuclear plant near the city.

There may be a way for residents to return soon after and stay on, however, and—in this scenario—it is achieved by the construction of domestic homes that are impervious to the radiation. These may be temporary emergency shelters, but they will soon become permanent features. Within each home, there are facilities to produce fresh water, grow adequate food crops, recycle waste, and be generally self-reliant. But to go outside and remain safe and healthy, it's necessary to wear a protective mask and suit. In effect, each home is like a little spaceship on the land, and indeed spaceship technology is likely to be used in the engineering and architecture of such residences.

Because Tokyo has been greatly depopulated, those who remain can create a quiet and independent postapocalyptic paradise. Some "Nuclear Greenies"—James Lovelock, for example—have noted that an irradiated landscape may in fact be a source of ecological recovery and rehabilitation, since most humans don't want to live there and abandonment of the radiated zone leaves wilderness alone to flourish (albeit with numerous mutative impacts and cancers). If such a conservation zone is paradise for animals and plants, it may also be paradise for a few intrepid Japanese families who enjoy the environmental challenges, the changed landscape, the wildlife, and the quiet isolation.

Tokyo 2121

□ □ □

Most of us do not like paying taxes, but in Toronto 2121, it's quite a fun activity. Every year the city government invites all registered Toronto residents to come to the annual participatory budgeting meeting, where everybody gets to decide on tax rates and where to spend the revenue. After a week of wrangling, the whole city then votes on which budget plan to follow.

Using this model of democratic budgeting, the citizens generally tend to opt for allocating financial resources not for roads and tax breaks or industrial development but for clean air projects, since Toronto has often suffered from a nasty summer smog problem. Voting for clean air also means voting for safe, pedestrianized streets as well as the development of many public parks and state-funded health programs.

Why do we assume this is what Torontoans would vote for? In a number of recent research studies, when a random selection of citizens is asked to identify the sort of policies they view as most worthy, these are the ones they usually cite. Of course, the untidy business of party politics, election campaigns, and an overrated thing called "leadership" usually messes up the way people vote in national elections, but at a city government level, in the forum of participatory budgeting, the policies rather than the personalities or the parties come to the fore.

In 2121, there are also two technological shifts that emerge from Toronto's annual budgeting session: (1) biogas energy plants are used, so that the city's waste—the household refuse and industrial waste—doesn't end up in landfills but instead is recycled to make energy, and (2) products that are not eco-friendly are subjected to higher tax rates. One upshot of this is that trucks are swapped for eco-drones, which, unencumbered by the weight of humans, are supposed to fly around speedily, without using roads or much fuel.

Toronto 2121

¤ ¤ ¤

Varna 2121 ⋈ Chronicle of the Sea Garden City

Varna, the marine capital of Bulgaria, is a city with a long and tortuous path to ecotopia:

4000 BC: Varna flourishes as a key trading site in the Copper Age on the western coast of the Black Sea.

600 BC–300 BC: Varna is refounded as a trading port by Greek merchants. The city is later claimed by Darius, the Persian emperor, before Alexander the Great takes it from him again for his own empire.

15 AD: Romans take the city, building palaces, roads, and bathhouses, all pleasantly overlooking the sea.

662–1444: Bulgarian Slavs move from somewhere east of the Black Sea to take Varna into their new kingdom. The Byzantines win it back. The Bulgarians take it back again. The Byzantines claim it from the Bulgarians once more, and so on and on for many centuries. Eventually, the Ottomans invade and colonize both Byzantium and Varna.

1854–1856: A combined Anglo-French army and navy uses Varna as the base camp for their campaign in the Crimean War against Russia. Despite huge numbers of allied troops dying of cholera, Britain and France go on to defeat Russia in the war.

1862: The Ottoman ruler of Varna commissions a small private garden by the sea.

1878: Russia attempts many times to "liberate" Varna from Ottoman rule, finally succeeding this year.

1881: The new Bulgarian mayor pays a French landscaper to greatly expand the "small private garden" into a "massive public garden" by the sea. Reports suggest corruption was involved in the arrangement.

1905: The sea garden grows to twenty-five acres, becoming the largest garden in the Balkans.

1946: Under Russian influence, Bulgaria, and Varna within it, submits to Communist Party rule.

1949: Varna is renamed "Stalin" after the Soviet leader (but reverts to Varna in 1956).

Varna 2121

◻ ◻ ◻

1989: Communism falls apart in Bulgaria, and the Varna leadership is democratically elected.

2006: Bulgaria becomes part of the European Union. Meanwhile, corrupt deals along the Varna coast are uncovered whereby politicians seem to pocket the proceeds of public lands sold to hotel developers.

2021: Most of the sea garden is privatized and sold off by Varna officials to luxury resort developers.

2121: Varna citizens stage an occupation of the old grounds of the sea garden, demanding that it be returned to the public and the hotels transformed into public schools.

Ecotopia 2121

¤ ¤ ¤

Vienna 2121 ¤ Agents of the Green Economy

Ecotopia 2121

¤ ¤ ¤

Wellington, a city on the North Island of New Zealand, faces a double whammy in the form of two significant future threats: human-induced climate change and a mighty natural disaster of a seismic nature.

One historical precedent for the latter threat is the Great 1855 Earthquake, which changed the coastal outline of Wellington forever, raising many square miles of new land from the sea. It was on this new land that the modern city was eventually built.

Before British colonization of the area, there were many such seismic events. Archeological evidence shows that Maori settlements repeatedly had to be abandoned in the Wellington region because earthquakes changed the layout of the coast, demolishing hunting and fishing zones that the indigenous settlements had depended on. Most geologists believe that the same sort of landscape-changing event is sure to recur sometime in the future, lifting great blocks of land up higher or lowering them down farther into the sea.

Accompanying these seismic shifts, Wellington's waterline will also probably change drastically because of sea-level rise, and this may mean the erosion and drowning of its low coastal areas by 2121. The eroded and drowned sections of the city will likely include the zones where the city's embassies, banks, and government buildings are currently located, along with the airport, the seaport, and part of the only highway that connects Wellington to the rest of the nation. During this epochal change, many Wellingtonians will probably make their way elsewhere, but many others will choose to stay.

To recover after such a tumultuous and isolating events, Wellington 2121 looks to the cultural history of the area. New Zealanders are proud of their agrarian heritage. After all, the nation became a very prosperous country early last century through farming. More recently, many New Zealanders have shown pride in their nation's association with Tolkien's *Lord of the Rings* world as represented in a series of blockbuster movies. This affection is such that, for a week each year, the Wellington council renames the city "Middle Earth," during which time various motifs and themes from the *Rings* movies adorn the city to make it look as though some pagan festival of the arts is in full bloom.

In Wellington 2121, a Shire-esque communitarian settlement is developed into a real-life extended village in the Wellington hills. Here, in this post-disaster zone, the central government has abandoned the remaining residents to let them run their own affairs, allowing them to harvest energy from Wellington's abundant winds and to trade in domestic agricultural produce—as in days of old.

In Wellington 2121, formal education has ground to a halt, but in its stead apprenticeships abound in the arts and crafts of aquaculture and boatbuilding. This enables young locals to develop an enviably productive and quietly profitable lifestyle. Those Wellingtonians seeking a more exciting way of life may then build for themselves a boat and sail freely to the South Island, which is visible in the distance.

According to some futurists, most modern societies are probably heading for new types of civilizations in the twenty-first century, and they are likely to get there through a cascading series of grand disruptions, as in the case of Wellington in this scenario. After such grand disruptions, cities may recover with a transformed sense of how to avoid similar crises in the future. We can imagine that Wellington's postdisaster future allows for such a transformation, and the new architecture of the city is one example. To survive earthquakes, architects opine that single-story buildings are the most durable and resilient. Thus, where now Wellington boasts more skyscrapers than any other city in the country, it will be distinctly "low-rise" come 2121 AD.

One other proud tradition in New Zealand that might be of use in Wellington 2121 is known as "Kiwi ingenuity." (New Zealanders have long referred to themselves colloquially as "Kiwis," a term adopted from the name of New Zealand's national bird, not from the name of the fruit, which also grows abundantly there). Kiwi ingenuity is a tradition of adaptability in the face of limited resources. The term usually reflects the ability of the average New Zealander to cobble together new devices from leftover bits of old machines. Generally, this is done in an ad hoc manner, without much pretense of finesse or replication. However, under the apprenticeship schemes in Wellington 2121, individual instances of Kiwi ingenuity are widely replicated among the community. From old bits of useless cars and computers, the community is able to fashion new and far more appropriate devices, such as small-scale wind turbines, plows, and the rigging for boats.

Wellington 2121

¤ ¤ ¤

All over the industrializing world, billions of people have moved into cities from rural settings. During the same time period, small villages have been transformed into large metropolises. This pattern is not followed here, however. Wellington 2121 will have transformed itself from what is usually perceived as "modern" to a "traditional" village-like settlement. Of course, if you ask local people from Wellington 2121 why they chose to "go back to the past," they may well appear confused by the question. This is not backward; this is forward. This is the way to survive and be happy.

Ecotopia 2121

¤ ¤ ¤

Wolverhampton 2121 ⌑ The Air That We Breathe

Now we journey from the idyllic shires of Middle Earth to its hellish deep interior: Mordor. This is the poisoned industrial land of Tolkien's *Lord of the Rings* novels, and it was inspired by a real place: Wolverhampton, a city in the Midlands of England. Wolverhampton was the first city of the Industrial Revolution. It was in Wolverhampton in the late eighteenth century that the steam engine was first put to industrial use, clearing mines of water. This was also where coal pits were first hacked into the city grounds to produce coal to serve the ever-growing ironworks and steelworks nearby. All these mines and factories sent dirty soot into the air and over the cityscape, earning Wolverhampton the unflattering nickname of the "Black Country"—which, incidentally, was also what the elves of *The Lord of the Rings* called Mordor.

Literary reflections of Wolverhampton were never very complimentary. Charles Dickens wrote that Wolverhampton's factories "[p]oured out their plague of smoke, obscured the light, and made foul the melancholy air." Even Queen Victoria, usually a paragon of polite expression, couldn't hide her feelings when she wrote in her diary of Wolverhampton as "a large and dirty town." Despite this bad press, many Wulfrunians are proud of their heritage, and they long for a time when the city can reassert itself as an industrial center.

So, in 2121, Wolverhampton's factories rise again, but this time via *eco-facturing*. In Wolverhampton 2121, blue-collar workers have been transformed into green-collar workers, and they have zero tolerance for air pollution. If any factory produces any form of smog, smoke, sooty cloud, greenhouse gas, or unsavory airborne smell, the factory is immediately shut down. For centuries, on cloudless days in the Wolverhampton of yesteryear, the sun shone to street level only through a fuzzy, reddish haze, such was the thickness of the dark smog. But in Wolverhampton 2121, the sunshine is bright and the air is finally clear.

For sure, this path to ecotopia is one-dimensional, for it singles out one particular physical factor and works to create an economy that revolves around that. However, the residents of Wolverhampton 2121 have argued and debated endlessly about possible paths toward Greening their city, from garden city plans to radical environmental education. Instead, they finally settle on a total cleanup of the one thing that just about all living beings depend upon minute by minute: the air that we breathe. Everything else can take care of itself.

Wuppertal 2121 ¤ Utopian Monorails

Ecotopia 2121

⬚ ⬚ ⬚

Where Wolverhampton was a pioneering industrial city for eighteenth-century England, Wuppertal was a pioneering industrial city for nineteenth-century Germany. Wuppertal has three particularly famous products from this period: aspirin, the world's longest functioning monorail, and the socialist theorist Friedrich Engels. This disparate group comes together in the rise of Wuppertal 2121.

When the private company that owns it announces the imminent closure of the aging monorail—planning to build parking lots in its place—many Wuppertal commuters grow very angry. The people of Wuppertal love their old monorail, saying it's an emblem of the city, that they can enjoy a fine, sunlit view of their environs when traveling, and that its use keeps automobile traffic in the streets to a minimum. However, the company says the monorail is just not economical.

So the citizens who support the monorail come together as a group, campaigning to "socialize" the monorail, dredging up a few of Engels's pamphlets as a guide. A deal is struck between the local council and monorail activists in which the council agrees to buy the monorail and all its assets if the monorail activists can get the money to pay for its upkeep and operation. Within days, ten-year subscription plans are put on sale and snapped up by eager commuters, who can travel for free for that length of time. It is such a successful endeavor that almost every citizen ends up with some kind of subscription. In fact, Wuppertal goes on to build more and more monorails of various styles all around the city. With so much commuting going on ten, twenty, fifty feet above street level, there's less desire for ground transport and more space opens up all around the city. Another feature popular with the locals is the construction of irregularly shaped buildings, circular buildings being the most common. These odd-shaped buildings allow for more open space in the city, since they leave small pockets of land that, due to their irregular dimensions, are completely unsuitable for private development save for small gardens or microbusinesses.

And what about the other signature product from Wuppertal, aspirin? Well, some overenthusiastic commuters sometimes need it for vertigo.

Xanadu 2121 ¤ Paradise Lost

Xanadu was the capital of the old Mongol empire. At its zenith in the thirteenth century, when Kublai Khan was emperor, it was home to at least one hundred thousand people, with maybe a million more traders, civil servants, and diplomats arriving and leaving over the course of the year. In the late fourteenth century, Xanadu was sacked and torched by the Ming army and then abandoned. For centuries, only a few temple ruins could be seen, along with some canal works. The city lived on and grew in mythic status through later literature, such as the poetry of Samuel Coleridge:

In Xanadu did Kubla Khan

A stately pleasure-dome decree:

Where Alph, the sacred river, ran

Through caverns measureless to man

Down to a sunless sea.

So twice five miles of fertile ground

With walls and towers were girdled round:

And here were gardens bright with sinuous rills,

Where blossomed many an incense-bearing tree;

And here were forests ancient as the hills,

Enfolding sunny spots of greenery.

Xanadu is now a UNESCO-protected site in China. Today, as in Kublai Khan's time, the area is mostly grassland, but year by year the rains become rarer. In the future, the site may be deserted arid wasteland. Every March, Yellow Dragons wreak havoc on the Xanadu site. These are violent sandstorms whipped up from the Gobi Desert that block out the sun for weeks at a time. Over the decades, they've become increasingly worse as the Gobi gets closer and ever more soil is denuded of grass and desiccated.

As China suffers environmental collapse during the tumultuous "dirty decade" sometime around the middle of the twenty-first century, the faltering Chinese government becomes fearful that Mongol tribes will try to reclaim Xanadu. Thus the site becomes heavily militarized. The Mongols never come, but by 2121 the Gobi Desert has fully invaded. Alas, what was once labeled a paradise is bound to end up as the cautionary opposite under China's current ecological trajectory.

Yerevan, the capital of Armenia, is a landlocked city in the Ararat plain between Mount Aragats and the Gegham mountain range. Though located in a broad valley, Yerevan is one of Eurasia's highest-altitude capitals. The air is not alpine fresh, though. The smoke from Yerevan's belching chimneys, rusty exhaust pipes, and numerous dump fires mixes with heavy, wet fog to choke the city with a smog so thick and rancid it can be tasted on the tongue and felt in the eyes. Many residents like to head to the mountains when they have a chance, to get away from the noxious fumes and into the fresh air. If they climb Mount Aragats, they might even get lucky and see Yerevan stretched out across the valley, but usually the smog hovering over the city is so thick that all you can see is a dirty yellow-brown mist and maybe a church steeple or two poking up above it.

The mountains surrounding Yerevan have become such an idyllic escape from the smoggy air that wealthy Yerevanis build their homes on their lower slopes, bribing local officials to encroach a little more year by year upon the forests of Mount Aragats. Apart from wealthy families and corrupt politicians, there are two other sections of Armenian society that inhabit the forests of the moun-

tain: orthodox monks and the Armenian army. The monks have been here for a thousand years or more, and the army since the mid-twentieth century, when Mount Aragats became a training ground. In the future, the numbers of both soldiers and orthodox monks have dwindled drastically as the fortunes of the church and the army have waned. In their stead, a wave of squatting hippies, New Age spiritualists, and university students comes and occupies the abandoned monasteries and soldiers' barracks, their small communes growing like mushrooms to propagate above the Yerevan smog.

Nobody knows whether the occupation is criminal or not, because these squatters are on land whose ownership is contested by the church and state, and no one has yet worked out who properly should be lodging a complaint with the police. Once in a while, an Armenian police cadet gets told to march up the mountain to investigate, but, of course, both the monastery and the army camp were built here precisely because the locale afforded them inaccessibility. So, after three hours of walking, the cadets usually give up, collect some delicious autumn mushrooms, then turn around and walk home again, leaving the squatters in peace.

Yiyang 2121 ✶ Last of the Dongting Dolphins

Yiyang is a city of five million people on the shores of the great lake of Dongting. Postcards of the city are resplendent with lovely, romantic scenes of this famous lake, and today it is a major domestic tourist attraction.

In the mid-twenty-first century, though, it is quite possible that, like the rest of China, Yiyang will experience environmental chaos that could kill many hundreds of thousands of people, leaving hundreds of thousands more homeless and millions jobless and bringing the Communist government to its knees because of the lack of resources and social unrest. Lake Dongting, the city's lifeblood, is likely to suffer worst of all:

- In 2060, large portions of the lake are filled in with landfill materials and reclaimed. While the city of Yiyang grows, the lake gets smaller.
- In 2061, the lake is deluged with contaminated mud, toxic sludge, burning oil spills, fiery floating garbage, pesticide residue, and fertilizers, meaning that the fish population plummets, decimating the fishing industry and the lake's biodiversity.
- In 2062, a series of episodic floods wipes out lakeside communities. The tourism

industry collapses because the image of Dongting Lake has changed from that of a pristine, watery paradise to a water-fiery hell.
- In 2063, plant and animal life seems to have all but disappeared from the lake, and Dongting is pronounced by ecologists to be biologically dead.

With free elections sprouting up all over China at about this time, it is apparent that two parties emerge in postcommunist Yiyang: the Blue Party and the Green Party. The Blue Party promises high GDP and high wages brought about by selling off all state industries on the "more efficient" free market. They also promise to implement new technologies to deal with the nasty ravages of nature. The Green Party promises to stop the war against nature and train every Chinese person how to make a living working *with* nature instead of *against* it. In Yiyang, in the wake of the Lake Dongting eco-crises, the Greener vision wins.

The path to ecotopia in Yiyang is symbolically tied to the fate of the Dongting freshwater dolphin. At the moment, the survival of this charming, smiling dolphin is precarious at best. Some hundred individuals are all that remain in the lake, which once teemed with tens of thousands. In 2064 they are extinct, a warning about the future to the humans in the area. But by 2121, they are miraculously alive once more. Some-how—probably from one of Lake Dongting's feeder rivers—a small population was able to survive the disasters of the mid-twenty-first century and then migrate back to the lake fifty years later. For Yiyang's Green Party, this is a symbol of their success as stewards of the city.

Yiyang 2121

¤ ¤ ¤

Zakynthos 2121 ⌘ Democratic Environmentalism

And so it's been rather a strange and winding voyage as we've zigzagged slowly around one hundred cities across seven continents around the globe. But here, now, we reach our final city of twenty-second-century ecotopia: Zakynthos, the capital of the Greek island of the same name. Zakynthos was the first place in the entire world to become an independent democracy, as written into its legal system some twenty-five hundred years ago. It kept this form of social organization intact for the remarkable period of six and a half centuries.

Today Zakynthos, situated in the sunny Ionian Sea, is famous as a tourist island, well-known for its seven thousand kinds of flowers and its nineteenth-century literary heritage. Both Dionysius Solomos, the national poet of Greece, and Ugo Fuscolo, the national poet of Italy, were born in Zakynthos and wrote their most famous works here.

Two millennia ago, though, Zakynthos was famous for something else: its tar. The great city-states of Greece, such as Athens and Sparta, would send their naval vessels to Zakynthos and beg to visit Lake Kerion on the island to extract the tar and smear it all over their ships. The tar served to insulate and preserve the ships, making them faster and more resistant to rot. But because Zakynthos was both independent and democratic, there were certain public protocols that all the Greek commanders had to adhere to before they got their hands on what they had come seeking. Leocrites, for example, the Athens navy commander, and his lieutenant, Mippus, might well have had to attend a public seminar in the city's vast amphitheater so as to take questions from the entire interested populace of the city. It would go something like this:

In front of twenty thousand people, Leocrites randomly picked out a little clay shard from a large ceramic pot with one person's name etched on it. The official pot attendant, himself randomly selected the day before, then announced the name for all to hear: "The first shard is that of Lasus! Please, Lasus, stand to ask Commander Leocrites your question."

Up in row H, in the north section, Lasus stood up, flicking his pastel blue tunic into place. The Zakynthons all looked over at him in anticipation of his question. "My question is this: why have you come here?"

Zakynthos 2121

☐ ☐ ☐

Leocrites stood up and yelled loudly, "We have come here to warn you that you will be attacked!"

To this, Lieutenant Mippus added, "And we have come here to ask you if we can have some lake tar."

"What for?" asked Lasus again.

"No, no," said the pot official aloud. "We've taken your one question." He pushed his hand into the pot to randomly pick out another clay shard. "Okay, the next shard belongs to Ogoros!"

"That's me!" yelled out a Zakynthon in the last row wearing a pastel yellow tunic. "So, who's going to attack us? Is it you?"

"No!" said Commander Leocrites, annoyed by the confusion. He then tried to explain that all they really wanted was the tar, but he didn't get very far.

"Yes, yes! That's enough," interrupted the pot official again. "Only one answer per question, please. Otherwise it becomes unfair," he said wagging a finger at the commander.

The commander was about to finish his point, but the official had sunk his hand into the pot and selected another shard. "Nicandos!" he announced.

A barefoot Zakyntho, dressed in a disheveled pastel pink tunic, stood up in the stands not far behind the commander, gazed into the distance, and, unsure of himself, belched out his question. "Umm, do you have mountains in Athens?"

"Yes," said Commander Leocrites, "of course!"

"Who cares about mountains?" said his lieutenant sharply. "We've come to tell you—"

But he was cut off by the pot official, once more chiding that they had a lot to get through and a yes-or-no answer would have to do for that particular question. The official reached down deep, one foot, two feet, three feet into the pot, and came out with a new shard and a new name. "Pallados!"

"Do your mountains go up or down?" came the question from Pallados. The lieutenant rolled his eyes.

"What are you talking about?" replied Commander Leocrites.

"Do they go up or down?" asked the Zakynthon from his seat in the amphitheater.

"Well . . ." said Leocrites, a little bewildered. "Up!"

"Up?" repeated the Zakynthon in astonishment, and the whole crowd appeared to echo his surprise.

"Yes! Up!" said the lieutenant forcefully. "The same way you'll be smashed when the Spartans arrive! Listen—"

"You are respectfully reminded," interrupted the pot official, "to answer only the question specifically addressed to you. Due democratic process must be adhered to." He then scooped up another shard. "Okay. Polyprax!"

An old Zakynthon wearing a pastel orange tunic stood up very slowly with the help of those around him. "Yes, yes, in case you didn't know," he said loudly and slowly, "our mountains go down. Sometimes. Usually, especially when you're standing on top of them. So it's quite exciting for us to hear about your mountains that go up."

"Ask your question please, Polyprax," said the official impatiently.

"Yes, okay," answered the old man. But when he spoke again, he admitted he had forgotten it. "Um . . . just a second . . . Let me think . . . Ummm . . ."

Lieutenant Mippus frowned at the slowness of the proceedings. Suddenly he stepped in front of the pot official and spoke aloud to warn the Zakynthons that they might soon be attacked by a fleet of crazy, heavily armed Spartans.

"Does that answer your question, Mr. Polyprax?" asked the official.

The old Zakynthon responded that he hadn't yet asked his question.

At this point, the official just reached for another shard of clay, saying, "Well, we have to move on—maybe next time."

"What next time?" yelled the old Zakynthon angrily. "There's twenty thousand of us here. What are the chances that I'll get picked again?"

The pot official interrupted, telling the old Zakynthon that he should have been more careful with his question.

"Here's my question, then," said the old Zakynthon, not giving up. "Who made you the pot official?"

At this, the entire stadium became tense, all eyes focusing on the official. The official indignantly waved his hand and announced that the question had not been authorized.

"Puh—you've become a pot tyrant!" said the old Zakynthon to the laughter and cheers of many in the crowd. Some of those nearest him started yelling out: "Pot tyrant! Pot tyrant! Pot tyrant!"

"Listen," Lieutenant Mippus started up again. "We just wanted to warn you! Now, may we collect some lake tar, please?"

"No, no, no, Lieutenant!" said the pot official, one hand flapping around violently and the other banging the pot. "You must wait before you can ask your questions. Now, where was I?"

"When will the attack begin?" yelled out an unknown Zakynthon from among the masses.

The indignant pot official demanded to know who had asked the question without his permission.

"It was me, Agoren!"

"Did I pull out your name?" asked the pot official as he fussed about with the clay shards in his hands.

"Shuddup about the pot names for a minute!" yelled out Agoren before proceeding to warn all present that they had better listen to the Athenians to find out more about the imminent attack.

The official, though, began loudly rattling off various rules of the democratic process.

"Damn the rules!" yelled Agoren excitedly. "What are we going to do?!"

"Yeah, what are we going to do?" shouted more Zakynthons in the rows nearby. Soon the whole east stand was on its feet yelling the same question in chorus. Another bunch of Zakynthons in the north section then started yelling for them to shut up and let the pot official do his job. Amid

the growing cacophony, the official screamed aloud the next name. "Larsus!"

"Hey—he's already had a question," yelled various Zakynthons from around the amphitheater.

"That was Lasus, not Larsus, you morons!" yelled another group. One Zakynthon then lurched into the air and crowd-surfed down to the center of the amphitheater, rushed to the pot, and kicked it over. It smashed to pieces, scattering thousands of clay shards on the ground. Cheers and jeers erupted in anarchic symphony.

Amid the chaos, Commander Leocrites shook his head and whispered to his lieutenant that they best leave to go find the tar for themselves.

⬚ ⬚ ⬚

What does this story of Zakynthos tell us? That democracy is slow? That democracy is clumsy? That democracy is too much talking and not enough action? Maybe, but Zakynthos's democracy lasted longer than any other in the world, and Lake Kerion is still surrounded by seven thousand flowers. What's good for Zakynthos in 500 BC might also be good for Zakynthos in 2121. Oh, and the Spartans never did invade the city.

Conclusion

Attitudes toward the utopian impulse in art and scholarship have shifted a lot since Thomas More published his book five hundred years ago. Sometimes utopianism was regarded as being an unproductive fantasy. Worse, it was thought dangerous because it takes focus away from social reality. Others see the utopian impulse as a practical step toward social change, if only by raising awareness of problems and deficiencies (and thus, perhaps, it is dangerous all the same to those who'd rather these problems and deficiencies remain hidden). My present attitude on this is that utopian thought is actually a very practical way of engaging the mind of any person who lives in a city and might have ideas about how they can improve it.

It seems to me that among the many leaders around the globe, varied as they are, most want their own people to believe that an ideal world is hopeless and impossible so that everyone gives up on any form of social activism and settles for the world of today. Hope and desire, mixed with a rich social imagination, can work together as potent antidotes to the complacency of accepting the status quo.

As a small sample of the power of imagination, and as a summary of the ecotopian investigations outlined in the preceding pages, let me give you two little exercises before you set this book aside.

First, imagine that by 2121 every one of the scenarios for the one hundred cities actually comes to pass. Take your time. Flip back through the various images to map out the cities where we've traveled. Then ask yourself: Is such profound, global change in our cities at all possible? Is it at all desirable?

Second, imagine that there is one, single city somewhere in the world in the year 2121 that exhibits in some form or another one feature from every one of the one hundred ecotopian

Conclusion

□ □ □

cities described in this book. Again, take a little time to flip back through the images, and then ask yourself: Is this possible? And if it is, is it desirable?

If you complete these exercises, I'm hoping you'll find a way to interpret your own journey through *Ecotopia 2121*. Needless to say, these exercises invite critique, and I'm not about to believe that all of the one hundred visions will be enjoyed or even tolerated by those seeking a Greener world. Some of the visions barely ask anything of us as citizens of the world beyond just the occasional tweak to policy or technology, while other visions ask us to become radical and revolutionary. Those who are fond of the latter are cautioned that such new worlds are unrealistic, impossible, or—wait for it—utopian. However, those who are fond of the former have to ask themselves whether they are merely making the world look Greener without really solving the environmental crises at all.

□ □ □

One argument against utopia that comes up again and again is that it is a type of totalitarian master planning, a grand vision dictated from on high in its totality, with all the answers already formed to cover every social problem. This is not at all my intention. I hope the reader will see, through the divergent and fragmented mosaic of scenarios presented here, with their diverse settings and allusions, that my own pretense toward utopianism has little to offer in the way of a unified grand plan. As I have an affinity for democracy, especially democracy in new, radical forms, I personally believe a grand "vision" or a "plan" should be put into effect only via robust democratic means. However, to be honest, I'm not really out to provide the blueprint for such a process. In this book, the power of utopianism lies in its propensity to highlight critical questions, rather than to provide a clear vision of an ideal society or a defined road map for how to get there.

And so, in this spirit of questioning, let me offer you an array of questions that arise as I attempt a summary of *Ecotopia 2121*. For those who seek more, there is, at the end, an "answer."

Are these one hundred varying expressions of ecotopia really meant to be earnest and serious, or are they mere satire and speculation? Are they asking us to identify a specific future for a specified city, or are they just pointing out the short-comings of human life today in order to warn us about where we are heading?

Don't these one hundred visions just conflate utopia with dystopia, and ecotopia with techno-topia? Or are these one hundred visions of the future attempting to show the diversity of the utopian imagination and, in the process, trying to undermine the idea that a single utopian plan can hope to gain consensus?

Are they suggesting some special, important relationship between the liberating impact of technology and the future? Or are they ambivalent and ambiguous about the supposed liberating effect of technology? Or maybe they are starkly negating the idea that environmental welfare can be improved via technology? Or do they, in the end, grudgingly admit that technology is going to win the day for those who manage to gain the skills to use it and control it?

Do these scenarios perhaps suggest that some cities are doomed in the face of the global environmental crisis—that they have no future—and that others are in need of radical action if they wish to survive to the next century?

What is the role of the history of cities when we contemplate their future? Do they offer useful narratives worthy of engagement? Isn't the whole point of dreaming about the future that it is a means to escape the past? Or is a city's history an inescapable part of its trajectory to the future?

I know this may seem rather mysterious and self-contradictory, but the answer to all of these questions is: *Yes!*

Conclusion

□ □ □

¤ ¤ ¤

Bibliography

The scenario for each city, as well as the introduction and conclusion sections, have drawn on studies and analysis by previous researchers, writers, and environmentalists. Listed here in alphabetic order is a small sampling of sources for each city that may be of help to those seeking more information.

Abu Dhabi 2121

Brunn, S. D., ed. *Engineering Earth: The Impact of Mega-Engineering Projects.* New York: Springer, 2011.

Davidson, C. M. *Abu Dhabi: Oil and Beyond.* New York: Columbia University Press, 2009.

Hoffman, A. "Dubai's Burj Khalifa." *Globe and Mail,* January 4, 2010, http://www.theglobeandmail.com/report-on-business/dubais-burj-khalifa-built-out-of-opulence-named-for-its-saviour/article1208413/.

Human Rights Watch. *Building Towers: Cheating Workers: Exploitation of Migrant Construction Workers in the United Arab Emirates.* New York: Human Rights Watch, 2006.

Sabban, R. *Maids Crossing: Domestic Workers in the UAE.* Saarbrucken: Lambert Academic Publishing, 2012.

Sanderson, E. W. *Terra Nova: The New World after Oil, Cars and Suburbs.* New York: Abrams, 2013.

Thomas, L. "Dubai Opens a Tower to Beat All." *New York Times,* January 4, 2010, http://www.nytimes.com/2010/01/05/business/global/05tower.html?_r=0.

Worth, K. D. *Peak Oil and the Second Great Depression.* Denver: Outskirts Press, 2010.

Accra 2121

Bang, J. M. *Ecovillages: A Practical Guide to Sustainable Communities.* Vancouver, BC: New Society Publishers, 2005.

Douglas, E., A. Kurshid, M. Maghenda, Y. McDonnel, L. Meclean, and J. Campbell. "Unjust Waters: Climate Change, Flooding and the Urban Poor in Africa." *Environment and Urbanization* 20:1 (2008): 87–97.

Hardoy, J., D. Mitlin, and D. Satterthwaite. *Environmental Problems in an Urbanizing World: Finding Solutions for Cities in Africa, Asia and Latin America.* London: Earthscan, 2001.

IGARSS '02. *IEEE International* 5 (2002): 2874–2876.

Magadza, C. H. D. "Climate Change Impacts and Human Settlements in Africa: Prospects for Adaptation." *Environmental Monitoring and Assessment* 61 (2000): 193–205.

Owusu, J. H. *Africa, Tropical Timber, Turfs, and Trade: Geographic Perspectives on Ghana's Timber Industry and Development.* Lanham, MD: Lexington Books, 2012.

Twumasi, Y. A., and R. Asomani-Boatang. "Mapping Seasonal Hazards for Flood Management in Accra, Ghana Using GIS." *Geoscience and Remote Sensing Symposium* (2002).

Almaty 2121

Alexander, C. "Soviet and Post-Soviet Planning in Almaty, Kazakhstan." *Critique of Anthropology* 27, no. 2 (2007): 165–181.

Bhavna, D. *Kazakhstan: Ethnicity, Language and Power.* London: Routledge, 2004.

Gossling, S., C. Michael-Hall, and D. Weaver, eds. *Sustainable*

Tourism Futures: Perspectives on Systems, Restructuring and Innovations. London: Routledge, 2012.

Feshback, M., and A. Friendly. *The Ecocide in the USSR: Health and Nature Under Siege.* New York: Basic Books, 1993.

Huttenbach, H. R. "Whither Kazakhstan? Changing Capitals: From Almaty to Aqmola/Astana." *Nationalities Papers* 26, no. 3 (1998): 581–587.

Janik, E. *Apple: A Global History.* London: Reaktion Books, 2011.

Kenessariyev, U., et al. "Human Health Cost of Air Pollution in Kazakhstan." *Journal of Environmental Protection* 4, no. 8 (2013): 869–876.

Luck, T. M. "The World's Dirtiest Cities." *Forbes Magazine,* February 26, 2008.

Magau, N., and R. Kaschek. "Remarks Concerning Traffic Problems of Almaty." *Central Asia Business Journal* 2 (2009).

Nabhan, G. P. "The Fatherland of Apples." *Orion,* May/June 2008.

Rasizade, G. "Russian Irredentism after the Georgian Blitzkrieg." *Contemporary Review,* March 22, 2009.

Sabol, S. *Russian Colonization and the Genesis of Kazakh National Consciousness.* New York: Palgrave MacMillan, 2003.

Sevcik, M. *Kazakhstan's Proposal to Initiate Commercial Imports of Radioactive Waste.* NTI Report, 2003.

White, S., and C. Moore. *Post-Soviet Politics.* London: Sage Publications, 2012.

Wolfel, R. L. "North to Astana: Nationalistic Motives for the Movement of the Kazakh(stani) Capital." *Nationalities Papers: The Journal of Nationalism and Ethnicity* 30, no. 3 (2002): 485–506.

Andorra la Vella 2121

Grassian, V. H., ed. *Nanoscience and Nanotechnology: Environmental and Health Impacts.* New York: Wiley, 2008.

Hunt, G., and M. Mehta. *Nanotechnology: Risk, Ethics and Law.* London: Earthscan, 2009.

Lead, G. R., and E. Smith, eds. *Environmental and Human Health Impacts of Nanotechnology.* New York: Wiley, 2009.

Vaccaro, I., and E. Beltran. *Social and Ecological History of the Pyrenees.* Walnut Creek, CA: Left Coast Press, 2010.

Antalya 2121

Campbell, K. M., R. J. Einhorn, and M. B. Reiss, eds. *The Nuclear Tipping Point.* Washington, DC: Brookings Institution, 2004.

Lucardie, P. *Democratic Extremism in Theory and Practice.* London: Routledge, 2013.

Martin, B. "Democracy without Elections." *Social Anarchism* 21: 18–51.

Pope, N., and H. Pope. *Turkey Unveiled.* London: Duckworth Publishing, 2011.

Pratt, J. A., M. V. Melosi, and K. A. Brosnan, eds. *Energy Capitals.* Pittsburgh, PA: University of Pittsburgh Press, 2014.

Scheer, H. *The Solar Economy.* London: Earthscan, 2002.

Athens 2121

Adkin, L. E. *The Politics of Sustainable Development: Citizens, Unions and the Corporation.* Montreal: Black Rose Books, 1999.

Copeland, B. R., and M. S. Taylor. *Trade and the Environment.* London: Routledge, 2005.

Forman, R. T. T. *Urban Ecology: The Science of Cities.* Cambridge, UK: Cambridge University Press, 2015.

Gossling, S., C. Michael-Hall, and D. Weaver, eds. *Sustainable Tourism Futures: Perspectives on Systems, Restructuring and Innovations.* London: Routledge, 2012.

Moussiopoulis, N. *Air Quality in Cities.* New York: Springer, 2003.

Waterfield, R. *Athens: From Ancient Ideal to Modern City.* New York: Basic Books, 2009.

Bibliography

¤ ¤ ¤

Bastia 2121

Bess, M. *The Light-Green Society: Ecology and Technological Modernity in France, 1960–2000.* Chicago: University of Chicago Press, 2003.

Blondel, M., J. Aronson, J. Y. Bodiou, and B. Boeuf. *The Mediterranean Region: Biological Diversity through Time and Space.* New York: Oxford University Press, 2010.

Boxer, B. "Mediterranean Pollution: Problem and Response." *Ocean Development & International Law* 10 (1982): 315–356.

Kousis, M., D. Ioannides, Y. Apostolopoulos, and S. Sonmez, eds. "Tourism and the Environment in Corsica, Sardinia, Sicily and Crete." In *Mediterranean Islands and Sustainable Tourism Development.* London: Cassell Academic, 2001.

Roberts, A. *Napoleon: A Life.* New York: Penguin Books, 2015.

Thornes, J. B., and J. Wainwright. *Environmental Issues in the Mediterranean.* London: Routledge, 2003.

Beijing 2121

Akabzaa, T. M. *Boom and Dislocation: the Environmental and Social Impacts of Mining in the Wassa West District of Ghana.* Accra: Third World Network Africa, 2001.

Chumley, C. A. "China Warns World: It's Time to 'de-Americanize.'" *Washington Times,* October 14, 2013.

de Lacerda, L., and W. Salomons. *Mercury from Gold and Silver Mining: A Chemical Time Bomb?* New York: Springer, 1997.

Economy, E. C. *The River Runs Black: The Environmental Challenge to China's Future.* Ithaca, NY: Council on Foreign Relations Books, Cornell University Press, 2010.

Goldhill, O. "China 'Has More Gold than Official Figures Show.'" *Telegraph Online,* April 8, 2014, http://www.telegraph.co.uk/finance/commodities/1075 3182/China-has-more-gold-than-official-figures-show.html.

Hiscock, G. *Earth Wars: The Battle for Global Resources.* New York: Wiley, 2012.

Hsu, M. Y. *Dreaming of Gold.* Stanford, CA: Stanford University Press, 2000.

Jain, R., J. Cui, and J. K. Domen. *Environmental Impact of Mining and Mineral Processing: Management, Monitoring, and Auditing Strategies.* Oxford, UK: Butterworth-Heinemann, 2013.

Mitchell, K. *Gold Wars.* Atlanta: Clarity Press, 2013.

Mourdoukoutas, P. "Can China Build a De-Americanized World?" *Forbes Magazine,* October 14, 2013.

Nriagu, J. O. "Mercury Pollution from the Past Mining of Gold and Silver in the Americas." *Science of the Total Environment* 149: 167–181.

Bethlehem 2121

Al-Khatib, I., and A. Eliewi. *Pollution Sources: Effect on the Water Environment and Livelihoods in Developing Countries: The Northern West Bank–Palestine.* Lambert Academic Publishing, 2010.

Gossling, S., C. Michael-Hall, and D. Weaver, eds. *Sustainable Tourism Futures: Perspectives on Systems, Restructuring and Innovations.* London: Routledge, 2012.

Kitto, J. *The History of Palestine.* Nysiros Publishers, 2014.

Raheb, M. *Bethlehem Besieged: Stories of Hope in Times of Trouble.* Ardmore, PA: Augsberg Fortress Press, 2004.

Tal, A. *Israel's Woodlands: From the Bible to the Present.* New Haven, CT: Yale University Press, 2013.

Birmingham 2121

Allen, R. C. *The British Industrial Revolution in Global Perspective.* Cambridge, UK: Cambridge University Press, 2009.

Edmonson, P., and S. Wells. *The Shakespeare Circle: An Alternative Biography.* Cambridge, UK: Cambridge University Press, 2015.

Fraser, T. *The Gunpowder Plot: Terror and Faith in 1605.* London: Weidenfeld & Nicolson, 2003.

Jones, P. *Industrial Enlightenment.* Manchester, UK: Manchester University Press, 2013.

Saunders, C. J. *The Forest of Medieval Romance.* Brewer Publications, 1993.

Shakespeare, W. *As You Like It.* Arden Shakespeare Series, 2004 (first published 1601).

Bibliography

❑ ❑ ❑

Bibliography

✕ ✕ ✕

Skipp, V. *Crisis and Development: An Ecological Case Study of the Forest of Arden 1570–1674.* Cambridge, UK: Cambridge University Press, 2008.

Teagle, W. G. *The Endless Village.* Nature Conservancy Council, 1978.

Bristol 2121

Allen, R. C. *The British Industrial Revolution in Global Perspective.* Cambridge, UK: Cambridge University Press, 2009.

Cristopher, J. *Brunel in Bristol.* Stroud, UK: Amberly, 2013.

Jones, P. *Industrial Enlightenment.* Manchester, UK: Manchester University Press, 2013.

Lyatkher, V. M. *Tidal Power.* New York: Wiley, 2013.

Budapest 2121

Cockrell-King, J. *Food and the City: Urban Agriculture and the New Food Revolution.* Amherst, NY: Prometheus Books, 2012.

Farnsworth, C. R. *Creating Food Futures.* Aldershot, UK: Gower, 2008.

Gille, Z. *From the Cult of Waste to the Trash Heap of History: The Politics of Waste in Socialist and Post-Socialist Hungary.* Bloomington, IN: Indiana University Press, 2007.

Mouget, L. J. A. *Growing Better Cities: Urban Agriculture for Sustainable Development.* Ottawa: IDRC Books, 2006.

Burlington 2121

Bookchin, M. *The Philosophy of Social Ecology.* Montreal: Black Rose Books, 1995.

David-Friedman, E. "Bernie Sanders and the Rainbow in Vermont." *Fire in the Hearth: The Radical Politics of Place in America* 4: 137.

Edmonson, B. *Ice Cream Social: The Struggle for the Soul of Ben and Jerry's.* Oakland, CA: Berrett-Koehler Publishers, 2014.

Eiglad, E. *Social Ecology and Social Change.* Porsgrunn, Norway: New Compass Press, 2014.

Mouget, L. J. A. *Growing Better Cities: Urban Agriculture for Sustainable Development.* Ottawa: IDRC Books, 2006.

Cape Town 2121

Bell, F. G., and L. J. Donnelly. *Mining and Its Impact on the Environment.* Boca Raton, FL: CRC Press, 2006.

Bodansky, D. *Nuclear Energy: Principles, Practices, and Prospects.* New York: Springer, 2008.

Schrader-Frechette, K. *What Will Work: Fighting Climate Change with Renewable Energy, Not Nuclear Power.* New York: Oxford University Press, 2012.

Scrase, I., and G. Mackerron, eds. *Energy for the Future: A New Agenda.* New York: Palgrave Macmillan, 2009.

Chicago 2121

Bodansky, D. *Nuclear Energy: Principles, Practices, and Prospects.* New York: Springer, 2008.

Brown, K. *Plutopia: Nuclear Families, Atomic Cities, and the Great Soviet and American Plutonium Disasters.* New York: Oxford University Press, 2015.

Clery, D. *A Piece of the Sun: The Quest for Fusion Energy.* London: Duckworth Publishing, 2013.

Groueff, S. *Manhattan Project: The Untold Story of the Making of the Atomic Bomb.* Bloomington, IN: iUniverse, 2000.

Spinney, R. *City of Big Shoulders: A History of Chicago.* DeKalb, IL: Northern Illinois University Press, 2000.

Chihuahua City 2121

Bell, F. G., and L. J. Donnelly. *Mining and Its Impact on the Environment.* Boca Raton, FL: CRC Press, 2006.

Bell, S., and S. Morse. *Sustainability Indicators: Measuring the Immeasurable.* London: Routledge, 2008.

Chales, G. *Cacti and Succulents: An Illustrated Guide to the Plants and Their Cultivation.* Marlborough, UK: Crowood Press, 2006.

Jain, R., J. Cui, and J. K. Domen. *Environmental Impact of Mining and Mineral Processing: Management, Monitoring, and Auditing Strategies.* Oxford, UK: Butterworth-Heinemann, 2015.

Mouget, L. J. A. *Growing Better Cities: Urban Agriculture for Sustainable Development.* Ottawa: IDRC Books, 2006.

Velducea, M. B. "Copper Mining in Chihuahua's 'Protected' Reserves." *Earth First Journal*, March 10, 2015.

Como 2121

Courtenay, E. "Clooney's Italian Lake Has Eco-Blues." *Treehugger*, August 1, 2007.

Gossling, S., C. Michael-Hall, and D. Weaver, eds. *Sustainable Tourism Futures: Perspectives on Systems, Restructuring and Innovations*. London, Routledge, 2012.

Masestti, E. *Lake Como*. Italian Itineraries, 2015.

Dawei City 2121

Charney, M. W. *A History of Modern Burma*. Cambridge, UK: Cambridge University Press, 2009.

Cocket, R. *Blood, Dreams and Gold: The Changing Face of Burma*. New Haven, CT: Yale University Press, 2015.

Ganesan, N., and K. Y. Hlaing. *Myanmar: State, Society and Ethnicity*. Singapore: Institute of South East Asian Studies, 2007.

International Commission of Jurists. *Dawei Special Economic Zone Should Protect Rights of Area Residents*, 2013.

Skidmore, M., and T. Wilson. *Myanmar: The State, the Community and the Environment*. Canberra: ANU Press, 2011.

Denver 2121

Billings, A. C. *Olympic Media: Inside the Biggest Show on Television*. London: Routledge, 2008.

Grierson, D. *Arcology and Arcosanti, Toward a Sustainable Built Environment, Global Environment: Policies and Problems*. London: Atlantic Books, 2007.

Karamichas, J. *The Olympic Games and the Environment*. New York: Palgrave Macmillan, 2013.

Lenskyj, H. J. *Olympic Industry Resistance: Challenging Olympic Power and Propaganda*. Albany, NY: SUNY Press, 2008.

Mangun, J. A., and M. Dyreson, eds. *Olympic Legacies: Intended and Unintended*. London: Routledge, 2013.

Preuss, H. *The Economics of Staging the Olympics*. Cheltenham, UK: Edward Elgar Publishing, 2006.

Shaw, C. A. *Five Ring Circus: Myths and Realities of the Olympic Games*. Vancouver, BC: New Society Publishers, 2008.

Soleri, P. *Arcology: The City in the Image of Man*. Paradise Valley, AZ: Cosanti Press, 2006.

Soleri, P., and J. Strohmeier. *The Urban Ideal*. Berkeley, CA: Berkeley Hills Books, 2001.

Dubai 2121

Allen, L. "Dark Side of the Dubai Dream." *BBC Panorama*, April 6, 2009.

Brunn, S. D., ed. *Engineering Earth: The Impact of Mega-Engineering Projects*. New York: Springer, 2011.

Fuchs, R. "Cities at Risk: Asia's Coastal Cities in an Age of Climate Change." *Analysis from the East West Center*, no. 96 (2010).

Karamichas, J. *The Olympic Games and the Environment*. New York: Palgrave Macmillan, 2013.

Salahuddin, B. "The Marine Environmental Impacts of Artificial Island Construction, Dubai." PhD diss., Duke University, 2006.

Shaw, C. A. *Five Ring Circus: Myths and Realities of the Olympic Games*. Vancouver, BC: New Society Publishers, 2008.

El Dorado 2121

Childress, D. H. *Lost Cities and Ancient Mysteries of South America*. Kempton, IL: Adventures Unlimited Press, 2015.

Rosen, B. *The Atlas of Lost Cities*. New York: Godsfield Press, 2008.

Silver, J. "The Myth of El Dorado." *History Workshop Journal* 34 (1992): 1–16.

Florence 2121

Franklin, D., ed. *Leonardo da Vinci, Michelangelo, and the Renaissance in Florence*. New Haven, CT: Yale University Press, 2005.

Friedman, G. *The Next 100 Years: A Forecast for the 21st Century*. New York: Anchor, 2010.

Kaehne, F. *Leonardo da Vinci: Dreams, Schemes and Flying Machines*. New York: Prestell Publishing, 2000.

Waller, J. *The Discovery of the Germ*. New York: Columbia University Press, 2003.

Gaia 2121

Antill, J. P. *Sofia Geography: Exploring Spirituality, Landscape and Archetypes*. Amberley, NZ: Harpagornis Publishing Limited, 2014.

Deane-Drummond, C. *Eco-Theology*. Winona, MN: Anselm Academic, 2008.

Marshall, A. *The Unity of Nature*. London: Imperial College Press, 2002.

Sahindou, I. *Hope for the Suffering Ecosystems of Our Planet*. New York: Peter Lang, 2014.

Goa 2121

Carrapatso, E., and E. Kurzinger. *Climate-Resilient Development: Participatory Solutions from Developing Countries*. London: Routledge, 2013.

Chhatre, A., and V. Saberwal. *Democratizing Nature: Politics, Conservation and Development in India*. India: Oxford University Press, 2006.

Cunningham, J. *Christianity and Nudity: A History*. Bloomington, IN: AuthorHouse, 2006.

Egger, L., and J. Egger. *The Complete Guide to Nudism and Naturism*. Wicked Books, 2009.

Gorham, K. and D. Leal. *Christianity and Naturism: Are They Compatible?* Cambridge, UK: Grove Books, 2000.

Ramana, M. V. *The Power of Promise: Examining Nuclear Energy in India*. India: Viking Press, 2012.

Gongshan 2121

Arthington, A. *Environmental Flows: Saving Rivers in the Third Millennium*. Berkeley, CA: University of California Press, 2012.

Gordon, F. D. *Freshwater Resources and Interstate Cooperation:* *Strategies to Mitigate an Environmental Risk*. Albany, NY: SUNY Press, 2009.

Salween Watch. *The Salween under Threat: Damming the Longest Free River in Southeast Asia*. Centre for Social Development Studies, 2004.

Shapiro, J. *China's Environmental Challenges*. Cambridge, UK: Polity Press, 2013.

Wang, P., and S. Dong. *The Large Dam Dilemma: An Exploration of the Impacts of Hydro Projects on People and the Environment in China*. New York: Springer, 2014.

Graz 2121

Gallagher, K. S. *The Globalization of Clean Technology*. Cambridge, MA: MIT Press, 2013.

Moore, J., and T. Shute. *The Hidden Cleantech Revolution*. Energy Publishers of America, 2010.

Motavalli, J. *Forward Drive: The Race to Build Clean Cars for the Future*. San Francisco: Sierra Club Books, 2001.

Pampanelli, A., N. Trivedi, and P. Found. *The Green Factory*. Boca Raton, FL: CRC Press, 2015.

Greenville 2121

Barden, L. S. "Recovery of Schweinitz's Sunflower, *Helianthus schweinitzii*, a Federally Listed Endangered Species, After Early and Mid-growing Season Controlled Burns in Piedmont North Carolina." *Bulletin of the Ecological Society of America* 75 (1994): 9.

Davis, E. D., N. A. Shinn, and D. Hood. "Rocky Microsites Prove Best for Establishing Schweinitz's Sunflower and Endangered Species of the Carolina Piedmont." *Ecological Restoration* 17 (1999): 171–172.

Guzowski, M. *Towards Zero-Energy Architecture: New Solar Design*. London: Laurence King Publishing, 2012.

Jordon, W. *The Sunflower Forest*. Berkeley, CA: University of California Press, 2012.

Bibliography

¤ ¤ ¤

Hanoi 2121

Adger, W. N., and P. M. Kelly. *Living with Environmental Change: Social Vulnerability, Adaptation and Resilience in Vietnam.* London: Routledge, 2001.

Carrapatso, E., and E. Kurzinger. *Climate-Resilient Development: Participatory Solutions from Developing Countries.* London: Routledge, 2013.

Jerneck, A., and L. Olsson. "Adaptation and the Poor: Development, Resilience and Transition." *Climate Policy* 8, no. 2 (2008): 170–182.

Mouget, L. J. A. *Growing Better Cities: Urban Agriculture for Sustainable Development.* Ottawa: IDRC Books, 2006.

Havana 2121

Bowman, M. *Cuba: Cars and Cigars.* Charleston, SC: Fonthill Media, 2013.

Carrapatso, E., and E. Kurzinger. *Climate-Resilient Development: Participatory Solutions from Developing Countries.* London: Routledge, 2013.

Clouse, C. *Farming Cuba: Urban Agriculture from the Ground Up.* New York: Princeton Architectural Press, 2014.

Mouget, L. J. A. *Growing Better Cities: Urban Agriculture for Sustainable Development.* Ottawa: IDRC Books, 2006.

Paul, A. "A Red and Green Utopia?" *NewMatilda.com,* February 4, 2010, https://newmatilda.com/2010/02/03/cuba-red-and-green-utopia/.

Houston 2121

Allenby, B. R. *Industrial Ecology: Policy Framework and Implementation.* Saddle River, NJ: Prentice-Hall, 1998.

Graedel, T. E. H., and B. R. Allenby. *Industrial Ecology and Sustainable Engineering.* Upper Saddle River, NJ: Pearson, 2009.

Marshall, A. *The Unity of Nature.* London: Imperial College Press, 2002.

Karachi 2121

Bredenoord, L., and P. van Linder, eds. *Affordable Housing in the Urban Global South: Sustainable Solutions.* London: Routledge, 2014.

Friedman, G. *The Next 100 Years: A Forecast for the 21st Century.* New York: Anchor, 2010.

Fuchs, R. "Cities at Risk: Asia's Coastal Cities in an Age of Climate Change." *Analysis from the East West Center,* no. 96 (2010).

Luck, T. M. "The World's Dirtiest Cities." *Forbes Magazine,* February 26, 2008.

Katun 2121

Freedman, E., and M. Neuzil. *Environmental Crises in Central Asia: From Steppes to Seas, From Deserts to Glaciers.* London: Routledge, 2015.

Li, X., and P. F. Shan, eds. *Ethnic China: Identity, Assimilation, Resistance.* Lanham, MD: Lexington Books, 2015.

Lozny, L. R. *Continuity and Change in Cultural Adaptation to Mountain Environments.* New York: Springer, 2015.

Mackerras, C. *China's Ethnic Minorities and Globalization.* London: Routledge, 2003.

Mikahilov, D. A. "Altai Nationalism and Archeology." *Anthropology and Archeology of Eurasia* 52, no. 2 (2013): 33–50.

Rossabi, M. *Governing China's Multiethnic Frontiers.* Seattle: University of Washington Press, 2004.

Wiegandt, E. *Mountains: Sources of Water, Sources of Knowledge.* New York: Springer, 2008.

Košice 2121

Arthington, A. *Environmental Flows: Saving Rivers in the Third Millennium.* Berkeley, CA: University of California Press, 2012.

Barlow, M., and T. Clarke. *Blue Gold: The Battle against Corporate Theft of the World's Water.* Toronto: McClelland & Stewart, 2010.

Fry, T. *Steel: A Design, Cultural and Ecological History.* London: Bloomsbury Academic, 2015.

Snadj, E., and K. Sivaramakrishnan. *Nature Protests: The End of Ecology in Slovakia.* Seattle: University of Washington Press.

icy in Russia. Cheltenham, UK: Edward Elgar Publishing, 2013.

Giriggs, S. *The Politics of Airport Expansion in the United Kingdom: Hegemony, Policy and the Rhetoric of "Sustainable Aviation."* Manchester, UK: Manchester University Press, 2013.

Iftimie, I. A. *Russian Natural Gas as an Instrument of State Power.* Washington, DC: Westphalia Press, 2015.

Mountain View 2121

Davies, P. *The Eerie Silence: Searching for Ourselves in the Universe.* New York: Penguin, 2011.

Marshall, A. "The Search for Extraterrestrial Us." *Australasian Science*, April 2000.

Resnik, D. B. *The Ethics of Science.* London: Routledge, 2005.

Moynaq 2121

Arthington, A. *Environmental Flows: Saving Rivers in the Third Millennium.* Berkeley, CA: University of California Press, 2012.

Bissel, T. *Chasing the Sea.* New York: Vintage, 2004.

Breckle, W. S., and W. Wucherer, eds. *Aralkum: A Manmade Desert.* New York: Springer, 2011.

Zonn, I. S., and M. Glantz. *The Aral Sea Encyclopedia.* New York: Springer, 2010.

Mumbai 2121

Boo, K. *Behind the Beautiful Forevers: Life, Death, and Hope in a Mumbai Undercity.* New York: Random House, 2014.

Norman, D. A. *The Design of Future Things.* New York: Basic Books, 2007.

Townsend, A. M. *Smart Cities: Big Data, Civic Hackers and the Quest for a New Utopia.* New York: W. W. Norton, 2014

Nador 2121

Montgomery, C. *Happy City: Transforming Our Lives through Urban Design.* New York: Penguin, 2015.

Mousdale, D. M. *Biofuels: Biotechnology, Chemistry and Sustainable Development.* London: Taylor and Francis, 2008.

Thornes, J. B., and J. Wainwright. *Environmental Issues in the Mediterranean.* London: Routledge, 2003.

Namibe 2121

Hecht, G. *Being Nuclear: Africans and the Global Uranium Trade.* Cambridge, MA: MIT Press, 2014.

Laity, J. J. *Deserts and Desert Environments.* New York: Wiley-Blackwell, 2008.

Padmalal, D., and K. Maya. *Sand Mining, Environmental Impacts: Collected Case Studies.* New York: Springer, 2014.

New Orenburg 2121

Brady, A. M., ed. *The Emerging Politics of Antarctica.* London: Routledge, 2014.

Hiscock, G. *Earth Wars: The Battle for Global Resources.* New York: Wiley, 2012.

Iftimie, I. A. *Russian Natural Gas as an Instrument of State Power.* Washington, DC: Westphalia Press, 2015.

Triggs, G. D. *The Antarctic Treaty Regime: Law, Environment and Resources.* Cambridge, UK: Cambridge University Press, 2009.

Worth, K. D. *Peak Oil and the Second Great Depression.* Denver: Outskirts Press.

New York 2121

Baker, L. *Built on Water: Floating Architecture and Design.* Salenstein, Switzerland: Braun Publishing, 2014.

Friedman, T. *Hot, Flat and Crowded: Why We Need a Green Revolution—And How it Can Renew America.* New York: Picador, 2008.

Montgomery, C. *Happy City: Transforming Our Lives through Urban Design.* New York: Penguin, 2015.

Seidman, M. *The Imaginary Revolution.* New York: Berghahn Books, 2004.

Washburn, J. *University Inc.: The Corporate Corruption of Private Education.* New York: Basic Books, 2006.

Bibliography

□ □ □

Nizhni Novgorod 2121

Cypher, J. *The Process of Economic Development*. London: Routledge, 2008.

Fanger, D. *Gorky's Tolstoy and Other Reminiscences*. New Haven, CT: Yale University Press, 2008.

Landry, C. *The Art of City Making*. London: Earthscan, 2006.

Siegelbaum, L. H. *Cars for Comrades: The Life of the Soviet Automobile*. Ithaca, NY: Cornell University Press, 2008.

Nuuk 2121

Conkling, E., and R. Alley. *The Fate of Greenland: Lessons from Abrupt Climate Change*. Cambridge, MA: MIT Press, 2013.

Emmerson, C. *The Future History of the Arctic*. New York: PublicAffairs, 2010.

Ordos City 2121

Beardson, T. *Stumbling Giant: The Threats to China's Future*. New Haven, CT: Yale University Press, 2013.

Bone, M., and D. Johnson. *Steppes: The Plants and Ecology of the World's Arid Regions*. Portland, OR: Timber Press, 2015.

Day, P. "Ordos: The Biggest Ghost Town in China." *BBC News Magazine*, March 12, 2012.

Li, X., and P. F. Shan, eds. *Ethnic China: Identity, Assimilation, Resistance*. Lanham, MD: Lexington Books, 2015.

Pearce, F. *When the Rivers Run Dry: What Happens When Our Water Runs Out?* London: Eden Project Books, 2007.

Rossabi, M. *Governing China's Multi-ethnic Frontiers*. Seattle: University of Washington Press, 2004.

Oxford 2121

Alvord, C. *Divorce Your Car*. Vancouver, BC: New Society Publishers, 2000.

Balish, C. *How to Live Well Without Owning a Car: Save Money, Breathe Easier, and Get More Mileage Out of Life*. Berkeley, CA: Ten Speed Press, 2006.

Crawford, J. M. *Car Free Cities*. International Books, 2002.

Freund, P. "Automobility and Its Discontents." *Capitalism Nature Socialism* 23, no. 3 (2012): 118–122.

Ghent, R. H., and A. Semlyen. *Cutting Your Car Use: Save Money, Be Healthy, Be Green*. Vancouver, BC: New Society Publishers, 2006.

Greer, J. M. *Not the Future We Ordered: The Psychology of Peak Oil and the Myth of Eternal Progress*. London: Karnac Books, 2013.

Montgomery, C. *Happy City: Transforming Our Lives through Urban Design*. New York: Penguin, 2015.

Sanderson, E. W. *Terra Nova: The New World after Oil, Cars and Suburbs*. New York: Abrams, 2013.

Palo Alto 2121

Evan, W. M., and M. Manion. *Minding the Machines: Preventing Technological Disasters*. Upper Saddle River, NJ: Prentice Hall, 2002.

Hansel, G. R., and W. Grassie. *H+/−: Transhumanism and Its Critics*. Bloomington, IN: Xlibris, 2011.

Lilley, S. *Transhumanism and Society*. New York: Springer, 2012.

Norman, D. A. *The Design of Future Things*. New York: Basic Books, 2007.

Segal, H. *Technological Utopianism in American Culture*. Syracuse, NY: Syracuse University Press, 2005.

Panama City 2121

Barros, C. "Trees Could Be the Ultimate in Green Power." *New Scientist*, September 10, 2009.

Beatley, T. *Biophilic Cities: Integrating Nature into Urban Design and Planning*. Washington, DC: Island Press, 2010.

Harmon, R., ed. "An Introduction to the Panama Canal Watershed." *The Rio Chagres* (2005): 19–27.

Keelart, S. R., and J. Heerwagen. *Biophilic Design*. New York: Wiley, 2008.

Sanderson, E. W. *Terra Nova: The New World after Oil, Cars and Suburbs*. New York: Abrams, 2013.

Bibliography

◻ ◻ ◻

Wilson, E. O. *Biophilia*. Cambridge, MA: Harvard University Press, 2015.

Paris 2121

Kosmodemyansky, A. *Konstantin Tsiolkovsky: His Life and Works*. Honolulu: University Press of the Pacific, 2000.

Maurer, E., and J. Richers. *Soviet Space Culture: Cosmic Enthusiasm in Socialist Societies*. New York: Palgrave Macmillan, 2011.

Noble, T. F. *The Religion of Technology: The Divinity of Man and the Spirit of Invention*. New York: Penguin, 1999.

Segal, H. *Technological Utopianism in American Culture*. Syracuse, NY: Syracuse University Press, 2005.

Wakeman, R. "Dreaming the New Atlantis: Science and the Planning of Technopolis in France." *Osiris, Journal of the History of Science Society* 18 (2003): 255–270.

Zabusky, S. E. *Launching Europe: An Ethnography of European Cooperation in Space Science*. Princeton, NJ: Princeton University Press, 1995.

Perth 2121.

Jackson, H., and K. Svenson. *Ecovillage Living: Restoring the Earth and Her People*. Cambridge, UK: Green Books, 2002.

Knoll, E-M., and P. Burger. *Camels in Asia and North Africa: Interdisciplinary Perspectives on Their Significance in Past and Present*. Vienna, Austria: Austrian Academy of Sciences Press, 2012.

Oreskes, N., and E. M. Conway. *The Collapse of Western Civilization: A View from the Future*. New York: Columbia University Press, 2009.

Pearce, F. *When the Rivers Run Dry: What Happens When Our Water Runs Out?* London: Eden Project Books, 2007.

Sanderson, E. W. *Terra Nova: The New World after Oil, Cars and Suburbs*. New York: Abrams, 2013.

Stein, M. *When Technology Fails: A Manual for Self-Reliance, Sustainability, and Surviving the Long Emergency.*

White River Junction, VT: Chelsea Green Publishing, 2008.

Philadelphia 2121

Blee, K. M. *Democracy in the Making*. New York City: Oxford University Press, 2013.

Douglass, I., and P. James. *Urban Ecology: An Introduction*. London: Routledge, 2014.

Forman, R. T. T. *Urban Ecology: The Science of Cities*. Cambridge, UK: Cambridge University Press, 2015.

Phnom Penh 2121

Bredenoord, L., and P. van Linder, eds. *Affordable Housing in the Urban Global South: Sustainable Solutions*. London: Routledge, 2014.

Brown, L. *Full Planet, Empty Plates: The New Geopolitics of Food Scarcity*. New York: W. W. Norton, 2012.

Brinkley, J. *Cambodia's Curse: The Modern History of a Troubled Land*. New York: PublicAffairs, 2012.

Bryld, E. "Potentials, Problems, and Policy Implications for Urban Agriculture in Developing Countries." *Agriculture and Human Values* 20, no. 1 (2003): 79–86.

Chandler, D. *A History of Cambodia*. Boulder, CO: Westview Press, 2007.

Dasgupta, S., U. Deichmann, C. Meisner, and D. Wheeler. *The Poverty/Environment Nexus in Cambodia and Lao People's Democratic Republic*. Policy Research Working Paper, World Bank, 2005.

Gottesman, E. R. *Cambodia after the Khmer Rouge: Inside the Politics of Nation Building*. New Haven, CT: Yale University Press, 2004.

Simone, A. "The Politics of the Possible: Making Urban Life in Phnom Penh." *Singapore Journal of Tropical Geography* 29, no. 2 (2008).

Sokha, C. "PM Blames Floods on Rubbish." *The Phnom Penh Post*, September 17, 2010.

Wingquist, G. *Cambodia Environmental and Climate Change*

Bibliography
¤ ¤ ¤

Policy Brief. Stockholm: University of Gothenburg and SEI Stockholm, 2009.

Pittsburgh 2121

Crawford, J. M. *Carfree Design Manual.* International Books, 2009.

Hall, K., and G. Porterfield. *Community by Design: New Urbanism for Suburbs and Small Communities.* New York: McGraw-Hill, 2001.

Johnston, S. A., and S. S. Nicholas. *The Guide to Greening Cities.* Washington, DC: Island Press, 2013.

Montgomery, C. *Happy City: Transforming Our Lives through Urban Design.* New York: Penguin, 2015.

Sanderson, E. W. *Terra Nova: The New World after Oil, Cars and Suburbs.* New York: Abrams, 2013.

Speck, J. *Walkable City: How Downtown Can Save America, One Step at a Time.* New York: Farrar, Straus and Giroux, 2012.

Plymouth 2121

Elliot. L. "Robots Threaten 15m UK Jobs." *Guardian*, November 12, 2015.

Ford, M. *The Rise of the Robots: Technology and the Threat of a Jobless Future.* New York: Basic Books, 2015.

Hobsbawm, E. *The Age of Revolution.* New York: Vintage, 1996.

Hooghe, L. *The Rise of Regional Authority.* London: Routledge, 2010.

Jones, S. *Against Technology: From the Luddites to Neo-Luddism.* London: Routledge, 2013.

Kaplan, J. *Humans Need Not Apply: A Guide to Wealth and Work in the Age of Artificial Intelligence.* New Haven, CT: Yale University Press, 2015.

Sale, K. *Rebels against the Future: The Luddites and Their War on the Industrial Revolution.* New York: Basic Books, 1996.

Whiteside, K. H. *Precautionary Politics: Principle and Practice in Confronting Environmental Risk.* Cambridge, MA: MIT Press, 2006.

Prague 2121

Gold, H. *Bohemia: Where Art, Angst, Love and Strong Coffee Meet.* Edinburg, VA: Axios Press, 2007.

Seigel, J. *Bohemian Paris.* Baltimore, MD: Johns Hopkins University Press, 1999.

Stover, L. *Bohemian Manifesto: A Field Guide to Living on the Edge.* New York: Bulfinch, 2004.

Tarnoff, B. *The Bohemians.* New York: Penguin, 2015.

Puno 2121

Carey, M. *In the Shadow of Melting Glaciers: Climate Change and Andean Society.* New York: Oxford University Press, 2010.

Cheshire, G., and H. Lloyd. *Peruvian Wildlife.* Chalfont St. Peter, UK: Bradt, 2008.

Foerster, B. *Lost Ancient Technology of Peru and Bolivia.* CreateSpace, 2013.

Stanish, C. *Lake Titicaca: Legend, Myth and Science.* Los Angeles: Cotsen Institute of Archaeology Press, 2011.

Thomson, H. *A Sacred Landscape: The Search for Ancient Peru.* New York: Overlook Press, 2008.

Wiegandt, E. *Mountains: Sources of Water, Sources of Knowledge.* New York: Springer, 2008.

Rekohu Te Whanga 2121

Holmer, M., and K. Black, eds. *Aquaculture in the Ecosystem.* New York: Springer, 2009.

King, M. *Moriori: A People Rediscovered.* New York: Viking, 1987.

King, M., and R. Morrison. *A Land Apart: The Chatham Islands of New Zealand.* London: Random Century Press, 1991.

McGuirk, J. *Radical Cities.* New York: Verso, 2014.

Miskelly, C., ed. *Chatham Islands: Heritage and Conservation.* Canterbury, NZ: Canterbury University Press, 2009.

Reno 2121

Cheek, M., and J. McNerny. *Clean Energy Nation.* New York: Amacom, 2011.

Bibliography

◻ ◻ ◻

Glassly, W. E. *Geothermal Energy: Renewable Energy and the Environment.* Boca Raton, FL: CRC Press, 2014.

Lloyd, D. B. *The Smart Guide to Geothermal: How to Harvest Earth's Free Energy for Heating and Cooling.* Masonville, CO: PixyJack Press, 2011.

Moe, A. W. *The Roots of Reno.* BookSurge, 2008.

Park, E. *The Happy, Fun, Party Travel Guide to Reno.* lulu.com, 2015.

Resistencia 2121

Arthington, A. *Environmental Flows: Saving Rivers in the Third Millennium.* Berkeley, CA: University of California Press, 2012.

Bartkowski, F. *Feminist Utopias.* Lincoln, NE: University of Nebraska Press, 1991.

Blee, K. M. *Democracy in the Making.* New York: Oxford University Press, 2012.

Little, J. *Feminist Philosophy and Science Fiction: Utopias and Dystopias.* Amherst, NY: Prometheus, 2007.

McGuirk, J. *Radical Cities.* New York: Verso, 2014.

Rio De Janeiro 2121

Arenas, F. *Utopias of Otherness: Nationhood and Subjectivity in Portugal and Brazil.* Minneapolis. MN: University of Minnesota Press, 2003.

Del Rio, V. "Urban Design and Conflicting City Images of Brazil: Rio de Janeiro and Curitiba." *Cities* 9, no. 4 (1992): 270–279.

McGuirk, J. *Radical Cities.* New York: Verso, 2014.

Tracey, D. *Urban Agriculture: Ideas and Designs for the New Food Revolution.* Vancouver, BC: New Society Publishers, 2011.

Vespucci, A. *The Letters of Amerigo Vespucci.* Translated by M. H. Markham. CreateSpace, 2012.

Rome 2121

Deane-Drummond, C. *Eco-Theology.* Winona, MN: Anselm Academic, 2008.

Francis I, Pope. *Encyclical on Climate Change and the Environment and Inequality.* New York: Melville House, 2015.

Horrell, D. G. *The Bible and the Environment.* London: Routledge, 2014.

Kaufman, S. R., and N. Braun. *Good News for All Creation: Vegetarianism as Christian Stewardship.* New York: Lantern Books, 2002.

Nash, J. A. "Ethical Concerns for the Global-Warming Debate." *Christian Century* 109, no. 25 (1992): 773–776.

Pullela, J. "Pope Francis Preparing Encyclical on Environment." *Huffington Post,* January 1, 2015.

Winright, T. L. *Green Discipleship.* Winona, MN: Anselm Academic, 2011.

Salto del Guiará 2121

Arthington, A. *Environmental Flows: Saving Rivers in the Third Millennium.* Berkeley, CA: University of California Press, 2012.

Gordon, F. D. *Freshwater Resources and Interstate Cooperation: Strategies to Mitigate an Environmental Risk.* Albany, NY: SUNY Press, 2009.

Iriondo, M. H., and J. C. Paggi. *The Middle Paraná River; Ecology of a Subtropical Wetland.* New York: Springer, 2007.

McGuirk, J. *Radical Cities.* New York: Verso, 2014.

San Diego 2121

Englander, J. *High Tide on Main Street: Rising Sea Level and the Coming Coastal Crisis.* Science Bookshelf, 2013.

Forman, R. T. T. *Urban Ecology: The Science of Cities.* Cambridge, UK: Cambridge University Press, 2015.

Zimmerman, T. *Submarine Technology for the 21st Century.* Bloomington, IL: Trafford Publishing, 2000.

San Francisco 2121

Beatley, T. *Green Urbanism: Learning from European Cities.* Washington, DC: Island Press, 1999.

Cockrell-King, J. *Food and the City: Urban Agriculture and the*

Bibliography
☐ ☐ ☐

New Food Revolution. Amherst, NY: Prometheus Books, 2012.

Duram, L. A. *Good Growing.* Lincoln, NE: Bison Books, 2005.

Forman, R. T. T. *Urban Ecology: The Science of Cities.* Cambridge, UK: Cambridge University Press, 2015.

Johnston, S. A., and S. S. Nicholas. *The Guide to Greening Cities.* Washington, DC: Island Press, 2013.

Lappe, F. M. *Diet for a Small Planet.* New York: Ballantine, 2010.

McGuirk, J. *Radical Cities.* New York: Verso, 2014.

Tarnoff, B. *The Bohemians.* New York: Penguin, 2015.

San Gimignano 2121

Kirschenmann, F. A. "Green vs. Gene." *Agricultural Technology* 73, no. 1 (1992): 32–37.

Mannion, M. *Frankenstein Foods.* New York: Welcome Rain Books, 2000.

McGuirk, J. *Radical Cities.* New York: Verso, 2014.

Morris, A. E. J. *History of Urban Form before the Industrial Revolution.* London: Longman, 1994.

Santiago 2121

Bell, F. G., and L. J. Donnelly. *Mining and Its Impact on the Environment.* Boca Raton, FL: CRC Press, 2006.

Bell, S., and S. Morse. *Sustainability Indicators: Measuring the Immeasurable.* London: Routledge, 2008.

LeCain, T. *Mass Destruction: The Men and Giant Mines That Wired America and Scarred the Planet.* New Brunswick, NJ: Rutgers University Press, 2009.

Tibbett, M., ed. *Mining in Ecologically Sensitive Landscapes.* Boca Raton, FL: CRC Press, 2015.

São Paulo 2121

Beatley, T. *Green Urbanism: Learning from European Cities.* Washington, DC: Island Press, 1999.

Cockrell-King, J. *Food and the City: Urban Agriculture and the New Food Revolution.* Amherst, NY: Prometheus Books, 2012.

Del Rio, V. "Urban Design and Conflicting City Images of Brazil: Rio de Janeiro and Curitiba." *Cities* 9, no. 4 (1992): 270–279.

Lenskyj, H. J. *Olympic Industry Resistance: Challenging Olympic Power and Propaganda.* Albany, NY: SUNY Press, 2008.

Tracey, D. *Urban Agriculture: Ideas and Designs for the New Food Revolution.* Vancouver, BC: New Society Publishers, 2011.

Watts, J. "Brazil Protests Erupt Over Bus Fares and World Cup Costs." *Guardian,* June 18, 2013.

Shanghai 2121

Brown, L. *Full Planet, Empty Plates: The New Geopolitics of Food Scarcity.* New York: W. W. Norton, 2012.

Caffaro, P., and E. Crist, eds. *Life on the Brink: Environmentalists Confront Overpopulation.* Athens, GA: University of Georgia Press, 2012.

Ching, Y. *As Normal As Possible: Negotiating Sexuality and Gender in Mainland China and Hong Kong.* Kong Kong: Hong Kong University Press, 2010.

De Roo, G., and D. Miller, eds. *Compact Cities and Sustainable Urban Development: A Critical Assessment of Policies and Plans from an International Perspective.* Farnham, UK: Ashgate Publishing, 2001.

Erturk, A., and D. J. Inman. *Piezoelectric Energy Harvesting.* New York: Wiley, 2011.

Grennalgh, S., and E. Winckler. *Governing China's Population: From Leninist to Neoliberal Biopolitics.* Palo Alto, CA: Stanford University Press, 2005.

Hawkes, G. *Sociology of Sex and Sexuality.* Maidenhead, UK: Open University Press, 1996.

Jeffreys, E. *Sex and Sexuality in China.* London: Routledge, 2009.

Mann, S. L. *Gender and Sexuality in Modern Chinese History.* Cambridge, UK: Cambridge University Press, 2011.

Pohlman, E. *Killing China's "One Child": Policy vs. Politics, Pollution vs. Population.* Planet Ethics Press, 2013.

Rimmerman, C. A., and K. D. Wald. *The Politics of Gay Rights.* Chicago: Chicago University Press, 2000.

Bibliography

▫ ▫ ▫

Scharping, T. *Birth Control in China 1949–2000: Population Policy and Demographic Development*. London: Routledge, 2002.

Sharjah 2121

Anderson, P. K. "The Behavior of the Dugong *(Dugong dugon)* in Relation to Conservation and Management." *Bulletin of Marine Science* 31 (1981): 640–647.

Brunn, S. D., ed. *Engineering Earth: The Impact of Mega-Engineering Projects*. New York: Springer, 2011.

Forman, R. T. T. *Urban Ecology: The Science of Cities*. Cambridge, UK: Cambridge University Press, 2015.

Fuchs, R. "Cities at Risk: Asia's Coastal Cities in an Age of Climate Change." *Analysis from the East West Center*, no. 96 (2010).

Karakiewicz, J. "The City and the Megastructure." In *Future Forms and Design for Sustainable Cities*, ed. M. Jenks and N. Dempsey. Milton Park, UK: Taylor and Francis, 2005.

Sinaia 2121

Gallagher, T. *Modern Romania: The End of Communism, the Failure of Democratic Reform, and the Theft of a Nation*. New York: New York University Press, 2008.

Hitchens, K. *A Concise History of Romania*. Cambridge, UK: Cambridge University Press, 2002.

Kligman, G. "On the Social Construction of 'Otherness': Identifying 'the Roma' in Post-socialist Communities." *Review of Sociology* 7, no. 2 (2001): 61–78.

Saul, N., and S. Tebbutt. *Role of the Romanies: Images and Counter Images of "Gypsies"/Romanies in European Cultures*. Liverpool, UK: Liverpool University Press, 2005.

Singapore 2121

Eckert, P. *Spaceflight Life Support and Biospherics*. New York: Springer, 2010.

Fuchs, R. "Cities at Risk: Asia's Coastal Cities in an Age of Climate Change." *Analysis from the East West Center*, no. 96 (2010).

George, C. "Consolidating Authoritarian Rule: Calibrated Coercion in Singapore." *Pacific Review* 20, no. 2 (2007): 127–145.

Intergovernmental Panel on Climate Change. *Synthesis Report: Contribution of Working Groups I, II, and III to the Fourth Assessment Report of the Intergovernmental Panel on Climate Change*. Cambridge, UK: Cambridge University Press, 2007.

Kameyama, Y., and A. P. Sari. *Climate Change in Asia: Perspectives on the Future Climatic Regime*. Tokyo: United Nations University Press, 2008.

Karakiewicz, J. "The City and the Megastructure." In *Future Forms and Design for Sustainable Cities*, ed. M. Jenks and N. Dempsey. Milton Park, UK: Taylor and Francis, 2005.

Matsuura, T., and R. Kawamura. *Water-Related Disasters, Climate Variability and Change: Results of Tropical Storms in East Asia*. Trivandrum, India: Transworld Research Network, 2007.

Mitchell, P. T. "Does Singapore's Destiny Lie in Outer Space?" *IDSS Commentaries* (59/2006), Nanyang University of Technology, Singapore.

Olds, K. *Globalization and Urban Change: Capital, Culture, and Pacific Rim Mega-Projects*. New York: Oxford University Press, 2002.

Turnbull, C. M. *A History of Modern Singapore*. Singapore: Singapore University Press, 2009.

Wei-Shiuen and R. Mendelsohn. "The Impact of Sea Level Rise on Singapore." *Environment and Development Economics* 10 (2005): 201–215.

Wong, B., and X. Huang. "Political Legitimacy in Singapore." *Politics and Policy* 38, no. 3 (2010): 523–543.

Sochi 2121

Dunbar, G. "FIFA Expels Chuck Blazer for Life for Bribery, Corruption." *USA Today*, July 9, 2015, http://www.usatoday.com/story/sports/soccer/2015/07/09/fifa-expels-chuck-blazer-for-life-for-bribery-corruption/29900353/.

Bibliography

◻ ◻ ◻

Gold, J. R., and M. M. Gold, eds. *Olympic Cities: City Agendas, Planning, and the World's Games, 1896–2016.* London: Routledge, 2010.

Hohman, S., and C. Mouradian. *Development in Central Asia and the Caucasus.* London: I. B. Tauris, 2014.

Karamichas, J. *The Olympic Games and the Environment.* New York: Palgrave Macmillan, 2013.

King, C. *The Ghost of Freedom: A History of the Caucasus.* New York: Oxford University Press, 2009.

Lenskyj, H. J. *Olympic Industry Resistance: Challenging Olympic Power and Propaganda.* Albany, NY: SUNY Press, 2008.

Longman, J. "Chuck Blazer, a Soccer Bon Vivant Laid Low." *New York Times,* May 27, 2015, http://www.nytimes.com/2015/05/28/sports/soccer/chuck-blazer-a-soccer-bon-vivant-laid-low.html?_r=0.

Socor, V. "Putin's Crimea Speech: A Manifesto of Greater-Russia Irredentism." *Eurasia Daily Monitor* 11, no. 56 (2014).

Stoeker, S., and R. Shakerova. *Environmental Crime and Corruption in Russia: Federal and Regional Perspectives.* London: Routledge, 2013.

Tanner, A. *The Forgotten Minorities of Eastern Europe: The History and Today of Selected Ethnic Groups in Five Countries.* East-West Books, 2004.

Springfield 2121

Beatley, T. *Green Urbanism: Learning from European Cities.* Washington, DC: Island Press, 1999.

Brown, K. *Plutopia: Nuclear Families, Atomic Cities, and the Great Soviet and American Plutonium Disasters.* New York: Oxford University Press, 2015.

de La Roca, C. "Matt Groening Reveals the Location of the Real Springfield." *Smithsonian Magazine,* May 2012.

Mugyenyi, B., and Y. Engler. *Stop Signs: Cars and Capitalism on the Road to Economic, Social and Ecological Decay.* Nova Scotia, Canada: Fernwood Publishing Company, 2011.

Owen, D. *Green Metropolis.* New York: Riverhead Books, 2010.

Schiller, P., and E. Brunn, eds. *An Introduction to Sustainable Transportation: Policy, Planning and Implementation.* London: Routledge, 2010.

Stuttgart 2121

Alvord, C. *Divorce Your Car.* Gabriola Island, Canada: New Society Publishers, 2000.

Balish, C. *How to Live Well Without Owning a Car: Save Money, Breathe Easier, and Get More Mileage Out of Life.* Berkeley, CA: Ten Speed Press, 2006.

Beatley, T. *Green Urbanism: Learning from European Cities.* Washington, DC: Island Press, 1999.

Goods, C. *Greening Auto Jobs: A Critical Analysis of the Green Job Solution.* Lanham, MD: Lexington Books, 2004.

Leitman, S., and B. Brant. *Build Your Own Electric Car.* New York: McGraw-Hill, 2013.

Mugyenyi, B., and Y. Engler. *Stop Signs: Cars and Capitalism on the Road to Economic, Social and Ecological Decay.* Nova Scotia, Canada: Fernwood Publishing Company, 2011.

Sloman, L. *Car Sick: Solutions for Our Car-Addicted Culture.* Cambridge, UK: Green Books, 2006.

Tracey, D. *Urban Agriculture: Ideas and Designs for the New Food Revolution.* Vancouver, BC: New Society Publishers, 2011.

Watkins, K. *Building an Electric Vehicle.* CreateSpace, 2012.

Worth, K. D. *Peak Oil and the Second Great Depression.* Denver: Outskirts Press, 2010.

Zachariades, T. *Cars and Carbon.* New York: Springer, 2011.

Sydney 2121

Beatley, T. *Green Urbanism: Learning from European Cities.* Washington, DC: Island Press, 1999.

Forman, R. T. T. *Urban Ecology: The Science of Cities.* Cambridge, UK: Cambridge University Press, 2015.

Mayer, B. *Blue-Green Coalitions.* Ithaca, NY: ILR Press, 2008.

Pearce, F. *When the Rivers Run Dry: What Happens When Our Water Runs Out?* London: Eden Project Books, 2007.

Bibliography

◻ ◻ ◻

Taipei 2121

Bracken, P. *Fire in the East: The Rise of Asian Military Power and the Second Nuclear Age.* New York: HarperCollins, 1999.

Grano, S. A. *Environmental Governance in Taiwan: A New Generation of Activists and Stakeholders.* London: Routledge, 2015.

Owen, D. *Green Metropolis.* New York: Riverhead Books, 2010.

Thimphu 2121

Ansari, M. *A Shangri-la Economy: Exploring Buddhist Bhutan.* Boca Raton, FL: Universal Publishing, 2012.

Brookes, J. S. "Buddhism, Economics, and Environmental Values: A Multilevel Analysis of Sustainable Development Efforts in Bhutan." *Society & Natural Resources: An International Journal* 24, no. 7 (2011): 637–655.

Evans, R. "The Perils of Being a Borderland People: On the Lhotshampas of Bhutan." *Contemporary South Asia* 18, no. 1 (2010): 25–42.

Dorji, T. "Sustainability of Tourism in Bhutan." *Journal of Bhutan Studies* 3 (2001): 84–104.

Giri, B. R. "Bhutan: Ethnic Policies in the Dragon Kingdom." *Asian Affairs* 35, no. 3 (2004): 353–364.

Gyamtsho, S. *Gross National Happiness and Social Progress: A Development Paradigm of Bhutan.* Lambert Academic Publishing, 2011.

Luechauer, D. L. "The False Promises of Bhutan's Gross National Happiness." *Global South Development Magazine,* July 21, 2013.

Mishra, V. "Bhutan Is No Shangri-La." *New York Times,* June 28, 2013.

Monaco, E. "Environmental Preservation and Development in the Kingdom of Bhutan: Utopia or Inspiration?" *Studies on Asia* 2, no. 1 (2012).

Rinzin, C. *On the Middle Path: The Social Basis for Sustainable Development in Bhutan.* Utrecht: Netherlands Geographical Studies, 2006.

Rizal, D. "The Unknown Refugee Crisis: Expulsion of the Ethnic Lhotshampas from Bhutan." *Asian Ethnicity* 5, no. 2 (2004): 151–177.

Thapa, S. J. "Bhutan's Hoax: Of Gross National Happiness." *Wave Magazine,* no. 3775, 2011.

Turner, M., S. Chuki, and J. Tshering. "Democratization by Decree: The Case of Bhutan." *Democratization* 18, no. 1 (2011): 184–21.

Walcott, S. M. "One of a Kind: Bhutan and the Modernity Challenge." *National Identities* 13, no. 3 (2011).

Wangchuck, A. D. W. *A Portrait of Bhutan.* New York: Penguin, 2006.

Werheim, J. *Bhutan: Hidden Lands of Happiness.* Chicago: Serindia Publications, 2006.

Timbuktu 2121

Boye, A. J., and J. O. Hunwick. *The Hidden Treasures of Timbuktu: Historic City of Islamic Africa.* London: Thames and Hudson, 2008.

Pearce, F. *When the Rivers Run Dry: What Happens When Our Water Runs Out?* London: Eden Project Books, 2007.

Stein, M. *When Technology Fails: A Manual for Self-Reliance, Sustainability, and Surviving the Long Emergency.* White River Junction, VT: Chelsea Green Publishing, 2008.

Tokyo 2121

Bortz, F. *Meltdown!: The Nuclear Disaster in Japan and Our Energy Future.* East Berlin, CT: 21st Century Publishing, 2012.

Elliot, R. *Fukushima: Impacts and Implications.* New York: Palgrave Macmillan, 2012.

IAEA. *Climate Change and Nuclear Power 2013.* Vienna, Austria: UN International Atomic Energy Agency, 2013.

Kameyama, Y., and A. P. Sari. *Climate Change in Asia: Perspectives on the Future Climatic Regime.* Tokyo: United Nations University Press, 2008.

Marcovici, M. *Lessons Learned: Nuclear Energy after Fukushima.* Books on Demand, 2013.

Martenson, C. *The Crash Course: The Unsustainable Future of our Economy, Energy and Environment.* New York: Wiley, 2011.

Mendell, W., ed. *Lunar Bases and Space Activities of the 21st Century.* Houston, TX: Lunar and Planetary Institute, 1985.

Salsberg, B., and C. Chandler. *Reimagining Japan: The Quest for a Future That Works.* San Francisco: VIZ Media, 2011.

Stein, M. *When Technology Fails: A Manual for Self-Reliance, Sustainability, and Surviving the Long Emergency.* White River Junction, VT: Chelsea Green Publishing, 2008.

Tucker, W. *Terrestrial Energy: How Nuclear Energy Will Lead the Green Revolution and End America's Energy Odyssey.* Austin, TX: Bartleby Press, 2012.

Toronto 2121

Beatley, T. *Green Urbanism: Learning from European Cities.* Washington, DC: Island Press, 1999.

Kresier, L,. and A. Sterling, eds. *Green Taxation and Environmental Sustainability.* Cheltenham, UK: Edward Elgar Publishing, 2012.

Lerner, J. A. *Making Democracy Fun.* Cambridge, MA: MIT Press, 2014.

Letcher, T. M. *Future Energy: Improved, Sustainable and Clean Options for Our Planet.* Philadelphia: Elsevier, 2008.

Milne, J. E., and M. S. Anderson. *Handbook of Research on Environmental Taxation.* Cheltenham, UK: Edward Elgar Publishing, 2015.

Owen, D. *Green Metropolis.* New York: Riverhead Books, 2010.

Varna 2121

Dicke, K., I. Dougherty, M. Jestice, P. Jorgensen, C. Pavkovic, and M. DeVries. *Battles of the Crusades 1097–1444: From Dorylaeum to Varna.* Gloucestershire, UK: Spellmount, 2007.

Forbes, N. *The Balkans: A History of Bulgaria and Serbia.* Pyrrhus Press, 2014.

Gounev, P., and V. Ruggiero. *Corruption and Organized Crime in Europe.* London: Routledge, 2014.

Vienna 2121

Beatley, T. *Green Urbanism: Learning from European Cities.* Washington, DC: Island Press, 1999.

Jones, V. *The Green Collar Economy.* New York: HarperOne, 2009.

Malkower, J. *Strategies for a Green Economy.* New York: McGraw-Hill, 2008.

McClelland, G. *Green Careers for Dummies.* Hoboken, NJ: Wiley, 2010.

Wellington 2121

Cari, D., S. Kindon, and K. Smith. "Tourists' Experiences of Film Locations: New Zealand as 'Middle Earth.'" *Tourism Geographies* 9, no. 1 (2007): 49–63.

Garcia, B. *Earthquake Architecture: New Construction Techniques for Earthquake Disaster Prevention.* Paco Asensio, 2011.

Holland, P. *Home in the Howling Wilderness: Settlers and the Environment in Southern New Zealand.* Auckland: Auckland University Press, 2013.

Jackson, H., and K. Svenson. *Ecovillage Living: Restoring the Earth and Her People.* Cambridge, UK: Green Books, 2002.

Jones, D., and K. Smith. "Middle Earth Meets New Zealand: Authenticity and Location in the Making of 'The Lord of the Rings.'" *Journal of Management Studies* 42, no. 3 (2005): 923–945.

Markley, O. "Manifesting Upside Recovery Instead of Downside Fear: Five Ways Mega-Crisis Anticipation Can Proactively Improve Futures Research and Social Policy." *Journal of Futures Studies* 16 (2011): 123–134.

Oreskes, N., and E. M. Conway. *The Collapse of Western Civilization: A View from the Future.* New York: Columbia University Press, 2014.

Sintubin, M., and I. S. Stewart, eds. *Ancient Earthquakes.* Boulder, CO: Geological Society of America, 2001.

Van Dissen, R., and K. Berryman. "It's Our Fault: Better Defining Earthquake Risk in Wellington." NZSEE 2009 Conference Proceedings.

Bibliography

¤ ¤ ¤

Wolverhampton 2121

Allen, R. C. *The British Industrial Revolution in Global Perspective*. Cambridge, UK: Cambridge University Press, 2009.

Braungart, M. *Cradle to Cradle: Remaking the Way We Make Things*. New York: North Point Press, 2002.

Cooper, C. D., and F. C. Alley. *Air Pollution Control: A Design Perspective*. Long Grove, IL: Waveland Press, 2010.

Hillis, D. R., and J. R. DuVall. *Improving Profitability through Green Manufacturing*. New York: Wiley, 2012.

Jones, P. *Industrial Enlightenment*. Manchester, UK: Manchester University Press, 2013.

Jones, V. *The Green Collar Economy*. New York: HarperOne, 2009.

McClelland, G. *Green Careers for Dummies*. Hoboken, NJ: Wiley, 2010.

Wuppertal 2121

Beatley, T. *Green Urbanism: Learning from European Cities*. Washington, DC: Island Press, 1999.

Engels, F. *Socialism: Utopian and Scientific*. 1880.

Groneck, C., and P. Lohkemper. *Urban Transport in Germany Pt. 9: The Wuppertal Suspension Railway*. Berlin: Robert Schwandl Verlag, 2007.

Pederson, C. *Monorails: Trains of the Future Now Arriving*. Monorail Society, 2015.

Xanadu 2121

Freedman, E., and M. Neuzil. *Environmental Crises in Central Asia: From Steppes to Seas, From Deserts to Glaciers*. London: Routledge, 2015.

Gallagher, S. *Meltdown: China's Environmental Crises*. Washington, DC: Pulitzer Center, 2013.

Ho, P., and E. Vermeer, eds. *China's Limits to Growth*. London: Routledge, 2006.

Man, J. *Xanadu: Marco Polo and Europe's Discovery of the East*. New York: Bantam, 2010.

Yerevan 2121

Harutyaunyun, A. *Yerevan and Its Neighbourhoods*. Turmanyan, Armenia: Edit Print Publishing House, 2012.

Hohman, S. *Development in Central Asia and the Caucasus*. London: I. B. Tauris, 2014

Yiyang 2121

Bell, S., and S. Morse. *Sustainability Indicators: Measuring the Immeasurable*. London: Routledge, 2008.

Gallagher, S. *Meltdown: China's Environmental Crises*. Washington, DC: Pulitzer Center, 2013.

Ho, P., and E. Vermeer, eds. *China's Limits to Growth*. London: Routledge, 2006.

Matsuura, T., and R. Kawamura. *Water-Related Disasters, Climate Variability and Change: Results of Tropical Storms in East Asia*. Trivandrum, India: Transworld Research Network, 2007.

Zakynthos 2121

Blee, K. M. *Democracy in the Making*. New York: Oxford University Press, 2013.

Lerner, J. A. *Making Democracy Fun*. Cambridge, MA: MIT Press, 2014.

Robinson, E. W. *Ancient Greek Democracy*. New York: Wiley-Blackwell, 2003.

Schofield, G. *Zakynthos: Complete Guide with Walks*. Lawrence, KS: Sunflowers Publishers, 2015.

Introduction and Conclusion

Anderson, E. N. *The Pursuit of Ecotopia: Lessons from Indigenous and Traditional Societies for the Human Ecology of Our Modern World*. Santa Barbara, CA: Praeger, 2010.

Appadurai, A. *The Future as Cultural Fact*. New York: Verso, 2013.

Callenbach, E. *Ecotopia*. Berkeley, CA: Banyan Tree Books, 1975.

Chew, S. C. *Ecological Futures*. Lanham, MD: Altamira Press, 2008.

Bibliography

¤ ¤ ¤

Cleays, G. *Searching for Utopia: The History of an Idea*. London: Thames and Hudson, 2011.

DeGeus, M. *Ecological Utopias: Envisioning the Sustainable Society*. International Books, 1999.

Fischer, M. J. *Anthropological Futures*. Durham, NC: Duke University Press, 2009.

Fishman, R. *Urban Utopias in the Twentieth Century: Ebenezer Howard, Frank Lloyd Wright, and Le Corbusier*. New York: Basic Books, 1977.

Friedman, G. *The Next 100 Years: A Forecast for the 21st Century*. New York: Anchor, 2010.

Glaeser, E. L. *The Triumph of the City*. New York: Penguin, 2011.

Jameson, F. *Archaeologies of the Future: The Desire Called Utopia and Other Science Fictions*. New York: Verso, 2005.

Lederwasch, A. "Scenario Art: A New Futures Method that Uses Art to Support Decision-Making for Sustainable Development." *Journal of Futures Studies* 17 (2012): 25–40

Marius, T. *Thomas More: A Biography*. Cambridge, MA: Harvard University Press, 1999.

Martenson, C. *The Crash Course: The Unsustainable Future of Our Economy, Energy and Environment*. New York: Wiley, 2011.

More, T. *Utopia*. New York: Penguin Books, 1982 (reprint edition).

Mumford, L. *The Story of Utopias*. New York: Viking, 1964.

Papanek, V. *Design for the Real World: Human Ecology and Social Change*. Chicago: Academy Chicago Publishers, 1984.

Spanos, C. *Real Utopia: Participatory Society for the 21st Century*. Chico, CA: AK Press, 2008.

Suvin, D. *Defined by a Hollow: Essays on Utopia, Science Fiction, and Political Epistemology*. New York: Peter Lang, 2010.

Tonkiss, F. *Cities by Design*. Cambridge, UK: Polity Press, 2013.

Bibliography

¤ ¤ ¤

¤ ¤ ¤

Acknowledgments

I'd like to dedicate this book to my three supervisors at Mahidol University: Wariya Chinwanno, Gamolporn Sonsri, and Wilasinee Anomasiri. Without their tolerance of my "foreign ways" and their faith in my scholarly abilities, this project would never have happened and this book would never have been written.

I wish to thank the following people who also provided the project with their support: Nanthawan Kaenkaew, Seree Surapong, Supaporn Songpracha, Sirirat Choonhaklai, Mark Felix, and the nonacademic staff on the Faculty of Social Sciences and Humanities at Mahidol University.

I must also acknowledge my colleagues and students in Mahidol University's Environmental Social Science program, who participated with me in the research and narrative (both artistic and scholarly) associated with the following cities: Kanang Kantamaturapoj (*Dawei City 2121*), Wannipol Mahaarcha (*La Paz 2121*), Wiporn Kanjanakaroon (*Nador 2121*), Thamakorn Siritorn (*Toronto 2121*), Patranit Srijuntrapun (*Hanoi 2121*), and Somnas Yanna (*Malé 2121*).

I would also like to acknowledge the help of Cal Barksdale and Amy Singh at Arcade Publishing and the contributors to the artwork for each city, as follows:

Abu Dhabi 2121: A. Marshall, Melkor 3D
Accra 2121: A. Marshall
Almaty 2121: A. Marshall
Andorra la Vella 2121: L. Jacobsen

Antalya 2121: A. Marshall
Athens 2121: A. Marshall, A. Dante
Bastia 2121: A. Marshall
Beijing 2121: A. Marshall, 3000AD
Bethlehem 2121: J. Grudzinski, D. Carillet, A. Marshall
Birmingham 2121: A. Marshall
Bristol 2121: A. Marshall, Iurrii
Budapest 2121: A. Marshall
Burlington 2121: A. Marshall
Cape Town 2121: Algol
Chicago 2121: A. Marshall
Chihuahua City 2121: A. Marshall, Melkor3D
Como 2121: Y. Gerzhedovich
Dawei City 2121: A. Marshall
Denver 2121: A. Marshall
Dubai 2121: A. Marshall, Melkor3D
El Dorado 2121: G. Nguyen, A. Marshall
Florence 2121: Melkor3D
Gaia 2121: J. Gerzhedovich, Cako, A. Marshall
Goa 2121: A. Marshall
Gongshan 2121: Firstear
Graz 2121: A. Marshall, J. Gerzhedovich
Greenville 2121: A. Marshall, G. Parisi
Hanoi 2121: A. Marshall
Havana 2121: A. Tsuper
Houston 2121: Inq
Karachi 2121: A. Marshall, K. Trutovsky
Katun 2121: A. Marshall, Melkor3D

Ecotopia 2121

¤ ¤ ¤

Košice 2121: A. Marshall

Lanzhou 2121: A. Marshall, N. Kaenkaew

La Paz 2121: Algol

Lazika 2121: A. Marshall

Leuven 2121: Unholy Vault Designs

London 2121: A. Marshall

Los Angeles 2121: A. Marshall

Macau 2121: A. Marshall

Madrid 2121: G. Nyuyen

Malaga 2121: A. Marshall

Malé 2121: Iurii

Mexico City 2121: A. Harburn, A. Marshall

Minsk 2121: A. Marshall

Moscow 2121: Crop

Mountain View 2121: Maodoltee

Moynaq 2121: SKDiz, A. Marshall

Mumbai 2121: E. Mendoza

Nador 2121: Unholy Vault Design

Namibe 2121: Isoga

New Orenburg 2121: A. Harburn, A. Marshall

New York 2121: Iurri, A. Marshall

Nizhni Novgorod 2121: A. Marshall

Nuuk 2121: A. Marshall

Ordos City 2121: A. Marshall, 3000AD

Oxford 2121: E. Mendoza

Palo Alto 2121: A. Marshall

Panama City 2121: E. Mendoza,

Paris 2121: Sdecoret

Perth 2121: A. Marshall

Philadelphia 2121: A. Marshall, Algol

Phnom Penh 2121: A. Marshall

Pittsburgh 2121: E. Mendoza

Plymouth 2121: D. Mcleod

Prague 2121: A. Marshall, L. Kulik

Puno 2121: Melkor3D

Rekohu Te Whanga 2121: A. Marshall

Reno 2121: Dark Geometry Studios

Resistencia 2121: S. Sunchoote

Rio de Janeiro 2121: A. Marshall

Rome 2121: A. Marshall, B. Rolff

Salto del Guairá 2121: A. Marshall

San Diego 2121: A. Marshall

San Francisco 2121: A. Marshall, L. Kulik

San Gimignano 2121: A. Marshall, PRILL, Unholy Vault Design

Santiago 2121: A. Marshall, N. Kaenkaew

São Paulo 2121: A. Marshall

Shanghai 2121: by E. Mendoza

Sharjah 2121: by Isoga

Sinaia 2121: A. Mit

Singapore 2121: A. Marshall

Sochi 2121: A. Marshall, Y. Gerzhedovich

Springfield 2121: Kanea

Stuttgart 2121: A. Marshall, N. Kaenkaew

Sydney 2121: A. Marshall

Tuipei 2121: A. Marshall

Thimphu 2121: E. Mendoza

Timbuktu 2121: PRILL

Tokyo 2121: A. Marshall

Toronto 2121: Georgeman

Varna 2121: A. Marshall

Vienna 2121: Diversepixel

Wellington 2121: A. Marshall

Wolverhampton 2121: A. Marshall

Wuppertal 2121: Denisgo

Xanadu 2121: Algol

Yerevan 2121: A. Marshall

Yiyang 2121: N. Maroz, A. Marshall

Zakynthos 2121: A. Marshall